International Construction Contract Management

International Construction Contract Management

D. Bryan Morgan

RIBA Enterprises

Published by RIBA Enterprises Ltd, 15 Bonhill Street, London EC2P 2EA

ISBN 1 85946 169 7

Stock Code 38404

British Library Cataloguing in Publications Data
A catalogue record for this book is available from the British Library

Publisher: Steven Cross
Commissioning Editor: Matthew Thompson
Project Editor: Anna Walters
Editor: David Hawthorn
Design and typesetting: Ben Millbank
Printed and bound by CPI Bath Press, Bath

Acknowledgements

We all have our favourite forms and procedures that we try to introduce into our current project. Many of these have developed over the years as we strive to improve, and eventually we tend to lose track of where the basic format originated. If any readers identify with any of the basic concepts behind any of the forms suggested in Appendices 3 through to 11 I trust that they will recognise this as an example of the evolution that is necessary in international construction and that they will not take offence, rather they will take it as a compliment. At no time have I intentionally set out to use material originally prepared by others.

Contents

Introduction	*page* 1
Acceleration – Directed and Constructive	7
Access to Site (Possession, Availability, Restrictions, Delays)	10
Alternative Dispute Resolution (ADR)	12
Ambiguities, Conflicts and Inconsistencies	19
Applicable Laws	21
Assignment and Novation	24
Bonds and Guarantees	29
Change Order Management	33
Claim Prompt List	38
Claim Review and Preparation	46
Claims and the Arbitrator	51
Concurrent Delay	54
Consideration	57
Contemporary Records	61
Contra Proferentem Rule	64
Contract Close Out	66
Contract – Maxims and Rules of Interpretation	69
Damages (including Consequential and Indirect Loss/Damage)	75
Dayworks	81
Decennial Liability	84
Delay Analysis	86
Depositions	91
Discovery	97
Document Control	102
Duty to Warn	105
Estoppel	109
Exclusive Remedies	112

Contents

Exemption Clauses	114
Fitness for Purpose	119
Float and its Ownership	122
Force Majeure	125
Fundamental Elements of Effective Contract Management	127
Implied Terms	137
Indemnity	141
Insolvency	144
Insurance Claims	149
Interrogatories	152
Kick-Off Meeting	155
Latent Defects – Limitation of Liability	159
Letters of Credit	162
Letters of Intent	165
Minutes of Meetings	171
Mitigation	174
Monthly Reports	177
Negotiation Procedures	181
Notices	186
Omission of Work – Loss of Contractor's Profit	189
Pre-Tender Information Report	191
Prevention	194
Quantum Meruit	197
Reasonable Skill and Care	201
Requests for Information (RFI Submissions)	203
Restitution	205
Restrictions on Trade	208
Retention	212

Contents

Rights of Third Parties	214
Risk Allocation, Assignment and Assumption	217
Ruling Language	223
Set-off	225
Subject to Contract	228
Subrogation	231
Sub-Soil Investigations Included with Tender Enquiry	233
Tender Rate Errors	235
Time at Large	238
Time for Review	242
Time of the Essence	244
Without Prejudice	247
Appendix 1 – Glossary of Terms	251
Appendix 2 – International Legal Systems	269
Appendix 3 – Contractor's Pre-Tender Site Review	279
Appendix 4 – Contractor's Pre-Tender Information Confirmation	289
Appendix 5 – Daywork Record Sheet	291
Appendix 6 – Overage, Shortage and Damage (OSD) Report	293
Appendix 7 – Monthly Progress Report Format	295
Appendix 8 – Change Order Proposal	299
Appendix 9 – Sub-Contractor Taking Over Certificate	301
Appendix 10 – Sub-Contract Close-Out Check List	303
Appendix 11 – Sub-Contractor Final Completion Certificate	305
Appendix 12 – Law Report Abbreviations & Table of Cases	307
Index	315

Introduction

Introduction

I have been involved in the construction industry in the UK and overseas for over 40 years in various contract, commercial and claims-related management and consultancy positions. During this time, and in particular during the 20 years or more which I have spent working on international construction contracts, I have become increasingly aware of the lack of a comprehensive, easy to understand, practical and 'down to earth' reference book for professionals involved in the management of international construction projects to guide them through the issues they face on a day to day basis. This book will hopefully be of use to architects, engineers, project managers, contract managers, claims consultants, quantity surveyors, contract administrators, contract engineers and anyone else involved in an aspect of management relating to a construction project.[1]

There is a wealth of books available discussing the standard forms of contract, including FIDIC; there are superb books on Liquidated Damages, although mostly directed at the standard JCT, GC/Wks/1, ACA, ICE and FIDIC contracts; there are Glossaries on INCOTERMS, Insurance Terms, International Trade Terms; there are masterpieces on Arbitration and Building and Civil Engineering Contract Law; there are manuals explaining Cost Control, Earned Value Management, Risk Management and Project Management techniques.

In practice, however, problems encountered while working on an international construction project seldom conform to the examples discussed in such books, superbly useful as they are in specific instances. Invariably the affected project is not proceeding under a FIDIC or other standard form of contract. In the majority of cases the contracts used for such projects are based upon either a hybrid version of FIDIC (if you are lucky) or, as is more often the case, a form of contract prepared by an engineer who is not a native English speaker and which has been prepared specifically for use on a particular project. The project you find yourself working on is not the four walls, floor and roof of a warehouse, which the referenced 'case studies' are often based upon, but a highly complex industrial plant located in the middle of nowhere where no-one has ever heard of JCT, ACE, FIDIC or even Hudson's Formula, and you are unable to find advice relating to anything near the type of situation you find yourself in.

This book provides many references to what has been established by case histories particularly in the UK, Hong Kong, Singapore, Australasia, South Africa, Canada and the US. I appreciate that these will not always be acceptable references in terms of the governing laws that apply to a particular contract where it is not based upon a Common law system but possibly a Civil Code system. They can still, however, serve as a general guide as to what is recognised in an international context as the basis upon which to resolve an otherwise obscure issue.

In a practical context, whilst possibly not providing any real support in a truly legal manner, such references may well conform to current interpretations under Civil Code rules or customary trade or industrial practices and can often provide extremely forceful arguments at the pre-litigation negotiation stage when parties are directing their efforts towards reaching an amicable solution to their dispute. Where possible I have tried to include comparisons with various Civil Codes, particularly those of France and Germany upon which many other countries have based their own Civil Codes.

Appendix 2 endeavours to identify which countries have legal systems based upon a Common law approach and which are based upon Civil Codes. It will be seen from Appendix 2 that several countries have adopted an amalgam of different legal systems upon which to base their own systems whereas other countries, whilst favouring a particular system in general terms, have accepted circumstances in which a different legal system can be utilised. From this it can be seen that the application of Common law case histories as discussed above can, in fact, have a very wide application throughout the world.

When interpreting contracts in foreign countries it is always wise to consult with a local lawyer[2] versed in commercial and construction contracts in order to compare the local laws, rules and regulations with those with which you are familiar. This book should be used as a guide and not as a definitive work on international contract law, which it is certainly not. It should direct the reader to issues that it would be beneficial to research in more detail as they relate to a particular project and be looked upon as a springboard to further knowledge.

I have intended this book to help the 'person in the field' who is facing issues that inevitably have to be resolved quickly and without outside assistance. When the issues get to the point where a higher level of assistance is needed I strongly recommend that a specialist contracts or claims consultant be appointed. If the basic principles suggested in this book have been followed, at least the consultant's job will have been made easier and the chances of success should have been improved.

No book will ever be able to cover every problem encountered on a construction project. In many ways, each project is a prototype since, even if the design is similar to one previously constructed, the odds are that yours will be in a different country and will be constructed under vastly different conditions. This book is intended to give both employers' and contractors' management personnel at best some guidance through the minefields in which we all work and at worst a second opinion on the issues they find themselves faced with. I hope that you find it helpful.

D. Bryan Morgan FRICS, MCIARB., FCMI, MAPM

April 2005

Introduction

1. There are different terminologies in use in different parts of the world for different jobs. For instance, a contract manager in the UK is usually traditionally recognized as the person managing the actual construction work for a contractor, possibly from an engineering or trade background. Not so in many countries overseas where this would be the title given to a quantity surveyor or engineer who is a specialist in contract forms, their preparation, interpretation and use.
2. A person that has been trained in the law and that has been certified to give legal advice or to represent others in litigation. Also known as a 'barrister' or 'solicitor'. In England and some other Commonwealth jurisdictions, a legal distinction is made between barristers and solicitors, the latter with exclusive privileges of advising clients, providing legal advice, and the former with exclusive privileges of appearing in a court on behalf of a client. Lawyers in some countries, such as Canada, sometimes use the title 'barrister and solicitor' even though, contrary to the position in England, there is no legal distinction between the advising and litigating roles. Canadian lawyers can litigate or give legal advice, as is the case in the USA, where lawyers are referred to as 'attorneys'.

International Construction Contract Management

Alphabetical reference

Acceleration – Directed and Constructive

An employer, faced with an excusable delay which could result in a claim for an extension of time from the contractor, may, either on its own initiative or on the suggestion of the contractor, consider it to be in the best interests of the contract to direct the contractor to accelerate the works in an attempt to meet the contract completion date. The employer is only able to do this, however, if there is a clause in the contract that gives it such authority. This type of acceleration is referred to as 'directed acceleration'.

If the employer considers this to be a favourable route to take it will ask the contractor for acceleration proposals in terms of time and cost and will either accept them, reject them or negotiate on them in the usual manner. The contractor is entitled to the costs of acceleration together with reasonable percentage additions for overheads and profit. In the final analysis, if the contractor takes the acceleration measures agreed upon yet still fails to meet the contract completion date, it will be entitled to the balance of the extension of time which it would otherwise have been entitled to had acceleration not been directed.

'Constructive acceleration' can be defined as a refusal by the employer to recognise in a timely manner that the contractor has encountered an excusable delay for which it is entitled under the contract to be granted an extension of time and which, because of the employer's failure to acknowledge this entitlement, results in the contractor having to accelerate its programme in order to attempt to complete the contract requirements by the existing completion date. This situation may arise either because the employer refuses a valid request for an extension of time or because it grants one at an inappropriate time.

In the US the doctrine of constructive acceleration is widely accepted, i.e. failure or refusal to grant a legitimate extension of time implies an instruction to accelerate. This legal doctrine has its origins in the US Court of Claims. It has been tried as an argument in England but has met with a certain amount of resistance. There is no such legal doctrine in England, nor indeed in most other Common law jurisdictions, and, in the absence of a specific clause relating to constructive acceleration, the best approach in such jurisdictions is probably to try to resolve this type of dispute using the contract dispute resolution mechanisms before considering the introduction of acceleration measures.

The opinion under the Common law system of England and Wales is that most construction contracts provide a remedy for failure to award an extension of time, this being the dispute resolution clause, whether the dispute is to be settled by the courts or by arbitration. It has been argued that the decision to accelerate was the contractor's, albeit probably to avoid liquidated damages, and was not the direct result of a breach by the employer. It has also been argued that constructive acceleration could be justified simply by applying the ordinary principles for breach of contract and damages to the implied instruction to accelerate which could be read into the situation. This implied instruction is said to bring about a collateral agreement between the parties in this respect. If this were not to be the case then the employer would benefit as a result of its own breach, which is contrary to the basic principles of law and so could not be allowed.

Despite Australia and Canada both being Common law jurisdictions similar to that of England and Wales, each have similar views to the US and two cases in point, *Perini Pacific* v *Commonwealth of Australia*[1] and *Morrison-Knudsen* v *BC Hydro and Power, Canada*[2] hinged on the facts that the engineer (Perini) in the first case and the employer (Morrison-Knudsen) in the second case each separately refused, or refused to consider, the contractor's requests for an extension of time. It was held in each case that the contractor had been instructed by implication to accelerate in order to maintain the original completion date. In each case the court held that the contractor was entitled to recover the costs incurred by it in accelerating.

It was held in the Singapore case of *Aoki Corporation* v *Lippoland (Singapore) Ltd*[3] that where the contract includes express timescales to be adhered to and where, as the result of an employer's action, those dates cannot be achieved, the contractor is entitled to recover the costs of additional resources necessary to accelerate the schedule as a result of such action by the employer. This is another example of a Common law jurisdiction founded on the Common law system of England and Wales finding favour with the predominantly US approach.

Whatever type of jurisdiction it is operating in, if a contractor makes a commercial decision to accelerate, it is always in its best interests to inform the employer of its intentions at the earliest opportunity and preferably before any acceleration measures are taken, emphasising that this decision has come about as a direct consequence of the employer's failure to grant an extension of time. In so doing the contractor is avoiding later disputes over the reasons for it having initiated acceleration measures and is also giving the employer the opportunity to confirm or deny its intention of paying the costs of such acceleration and/or to monitor and keep contemporaneous records of whatever acceleration measures the contractor takes.

The loss suffered as a result of the acceleration must be clearly shown as having arisen directly from the acceleration, either through the employer's breach of contract or through its implied instruction to accelerate. This exercise relies heavily upon the availability of accurate programmes, both anticipated and actual, from which anticipated resources can be compared against the actual resources expended. It also requires detailed record keeping sufficient to justify the additional resources claimed.

For a claim for constructive acceleration to be successful a contractor would need to show that it had a legitimate claim for delay which had either been rejected, not granted within the stipulated time period or not even considered by the employer, despite the contractor having served all the correct notices at the correct times in accordance with the requirements of the contract. The contractor's chances of the claim succeeding would also benefit from it having received some direction from the employer to proceed regularly to meet the contract completion date or even to employ additional resources to do so. Care must be taken in maintaining records to ensure that there is no overlapping of costs and that only those costs directly attributable to the additional acceleration-related resources are claimed.

In summary:

Directed acceleration:

- proposed by employer either on its own initiative or at the suggestion of the contractor;
- requires a specific clause in the contract giving the employer such authority (irrespective of what country or jurisdiction relates to the contract);
- the specific clause should confirm that, in the case of directed acceleration, the contractor shall be entitled to recover all costs (including reasonable percentages for overheads and profit) incurred by it in complying with the acceleration direction;
- the specific clause should confirm that if the contractor reduces but does not extinguish the delays to the time for completion for which it would otherwise have been entitled to an extension of time, then the contractor will be entitled to an extension of time for the balance of the delay period remaining.

Constructive acceleration:

- acceleration imposed upon the contractor by the employer refusing to grant an extension of time in relation to an excusable delay and holding the contractor to the date for completion included in the contract;
- England and Wales – hesitant to adopt the principle of constructive acceleration;
- US – a widely accepted principle adopted in the negotiation and settlement of construction claims;
- Australia – a similar approach to the US;
- Canada – a similar approach to the US;
- Singapore – a similar approach to the US;
- Civil law jurisdictions – typically more concerned with rules rather than precedents, hence application of the principle of constructive acceleration would depend upon the wording of the contract. It would be wise to cover this eventuality by including a clause in the contract defining if, when and how constructive acceleration would be recognised and dealt with.

1. *Perini Corporation* v *Commonwealth of Australia* (1969) 12 BLR 82.
2. *Morrison-Knudsen Co. Inc. and Others* v *British Columbia Hydro and Power Authority* (1991) 7 Const. LJ 227.
3. *Aoki Corp.* v *Lippoland (Singapore) Pte. Ltd* (1995) 2 SLR 609.

Access to Site (Possession, Availability, Restrictions, Delays)

Various issues come to mind when considering possession, availability, restrictions and delays in respect of the contractor being afforded proper access to the project site. The contractor requires access to the project site in order that it may commence its works and frequently in international contracts the date from which the contract period runs is the date when possession of the site occurs. Possession in turn may depend upon the various project finance packages becoming effective, usually where the contractor is from a different country from that in which the employer/project is located and has agreed in its tender bid to arrange for export credit financing or some other form of international finance. Possession may also depend upon the contractor depositing the requisite bonds, parent company guarantees and insurances with the employer.

Where the contract states that work is to be completed within a stipulated time and in a stipulated sequence, it is an implied term of the contract that the contractor will be given possession of the site immediately all the pre-requisites for possession have been achieved. In the case of *Freeman* v *Hensler*[1] the plaintiff contracted to demolish 15 houses and build 12 new houses in their place, all within six months of the date of the contract. The plaintiff then agreed to a two weeks delay in both the dates of possession and completion of the site.

It transpired that actual possession was not given until long after the agreed two weeks delay had passed and even then only partial possession was given, the site being handed over piecemeal thereafter. The works should have been largely completed within summer months but were delayed well into the winter. It was held that the contract included an implied term that possession would be within a reasonable time of the date of the contract. Despite the plaintiff agreeing to a two weeks delay to the dates for possession and completion, it was held that a reasonable time did not extend beyond that two weeks and the plaintiff was held entitled to damages for the loss he had sustained by reason of the delay.

In an earlier case[2] it was noted that a contractor will require a sufficient degree of possession of the site to enable it to carry out its works and this must be considered as an implied pre-condition of the contract. It must also be implied that such possession will be afforded within a reasonable time of signing the contract. During the contract the contractor will require sufficient uninterrupted possession to be able to execute the works through to completion. In the case of *Wells* v *Army & Navy*[3] it was noted that where a contract specifies a set time within which the contract works have to be completed, not only does the contractor have an obligation to complete within that time, there is also an obligation on the part of the employer such that the contractor is to be afforded that time within which to execute those works. A delay in possession or interruptions in unrestricted access during the progress of the works would fall foul of this latter obligation and leave the employer open for a claim from the contractor.

Generally on new construction contracts, unless otherwise stated in the contract, the employer must give possession of the whole site on the date of possession or commencement of the contract (as opposed to contracts for the alterations of existing structures where total access might necessarily be restricted) or else it could be held to be in breach of contract. Where the employer only gives possession of part of the site, the contractor must consider whether it needs the whole site in order to make a start or whether it can start with what it has been given. If the latter is the case then the contractor should make such a start, otherwise it could prejudice later claims for time and costs due to its failure to mitigate the effects of the restriction of access. If the contractor cannot make a start or if it makes only a limited start it will in both cases, unless specifically excluded by the contract, have a claim for additional time and costs.

During the contract there will be occasions when the employer or the employer's representatives or agents will require access either to the construction site or to the place of manufacture of any equipment to be incorporated in the work. Such access might be required to view progress or to carry out quality inspections. Procedures to be adopted for quality inspections and tests will usually be covered by a specific clause in the contract; however, there should in any case also be a separate clause in the contract setting out the obligations of the parties regarding the employer's rights of access to the work, particularly in respect of the employer's obligations to abide by the contractor's particular site health and safety regulations[4] in addition to all relevant statutory regulations, bye-laws and so on of the country.

Similarly, during the defects liability period the contractor will require access to the works already handed over in order to carry out maintenance inspections, remedial works and so on. Such access will have to be at times acceptable to the employer but restrictions in this respect must not be unreasonable. A clause should be included in every contract to ensure that all parties are aware of their rights, obligations and responsibilities in this respect.

There are no significant differences in the approaches to site access rights that occur purely because those projects are located in different countries or under different jurisdictions. The main reasons for different approaches usually lie in the wording of the relevant clauses used in each particular contract and it is believed that the experiences noted from Common law cases provide a fair and reasonable basis for negotiation in most jurisdictions.

1. *Freeman & Son* v *Hensler CA* (1900) 64 JP 260.
2. *Roberts* v *Bury Commissioners* (1870) LR 5 CP 310 at pp.320 and 325.
3. *Wells* v *Army & Navy Co-operative Society* (1902) 86 LT 764.
4. These will normally be spelt out in the contract under the appropriate clause headings including Health and Safety, Safety of the Works, Legislation, Rules and Regulations, or similar.

Alternative Dispute Resolution (ADR)

Alternative Dispute Resolution (ADR) is a term that covers a variety of different ways of seeking to facilitate the resolution of disputes without recourse to the courts and can involve different processes and can be triggered by different events in different countries. There are various ADR mechanisms available including mediation, conciliation, adjudication, expert determination, early neutral evaluation and mini-trials. In the context of this section, arbitration is excluded as an ADR mechanism because although it is technically an alternative to litigation it has many characteristics that are similar to litigation, making it a parallel process rather than an alternative one.

ADR has several advantages over litigation as a means of resolving construction disputes. ADR is considered to be:

- less expensive than litigation/arbitration;
- potentially quicker than litigation/arbitration;
- kinder to the long-term relationships between the parties;
- a means of enhancing communication between the parties;
- as private as the parties desire it to be;
- non-binding upon the parties until a settlement is agreed;
- a means to preserve the parties' legal rights and available processes until a final settlement is agreed;
- concerned with the settlement of the dispute only;
- an option that is available at any stage of the contract or the dispute;
- a procedure which enables the parties' principals to participate to a greater extent in the early resolution of the dispute.

On the minus side, ADR could be said to have the following disadvantages:

- a final solution is not a guaranteed end result as either party can abandon the mechanism at any time;
- a higher degree of openness than a party may be prepared to conform to may be necessary in order to progress the process;
- the end result does not provide a precedent which can be applied by third parties in similar disputes;
- each party must consent to the process and must cooperate fully in the process in a consensual rather than an adversarial scenario.

It is worth noting that the following explanations of the various terms used in ADR are those most commonly found in international construction. They are the explanations favoured by the ICC in relation to the various ADR services that they provide. It is possible to come across different interpretations of some of these terms, as an example the London Court of International Arbitration is of the opinion that 'conciliation' is merely 'mediation' by another name, although they recognise that others argue that the two involve different techniques.[1] They also make the valid point that in the US, ADR is understood to include arbitration.[2]

Mediation

Mediation, probably the most widely known form of ADR, is the mechanism whereby a third party acts as a facilitator to encourage discussion and negotiation between the parties in an attempt to reach a mutually acceptable compromise on the dispute referred to mediation. The third party mediator might be an individual or a mediation panel, in either case here referred to as the mediator. The role of the mediator is to listen to the cases put forward by the parties and to guide their negotiations towards a settlement, offering advice in the process as deemed appropriate at the time whilst maintaining neutrality. The mediator cannot impose a decision on either party and merely acts as a channel through which the parties can negotiate.

It is important for the parties to select an appropriate mediator taking into consideration the subject matter and size of the dispute and the location of the parties to the dispute. They can either organise the appointment and the mediation process themselves or appoint one of the readily available mediation organisations to do this on their behalf.[3]

The mediation process usually involves company representatives of the parties familiar with the dispute, and with the authority to agree a solution to the dispute, who are accompanied by legal advisers. Each party usually makes an opening statement after which they separate and the mediator takes turns to discuss the case with each party, identifying common ground and guiding them towards a compromise solution. The mediator assists the parties in reaching a mutually acceptable resolution to their dispute rather than rendering a decision. The negotiations are 'without prejudice' and cannot be referred to in any subsequent litigation process.

Conciliation

The conciliation concept and the conciliation process are very similar to mediation in many ways, the main difference in its application by such bodies as the ICC being that in conciliation the conciliator takes a more active role in finding a solution to the dispute and is permitted to make suggestions on a possible compromise rather than merely acting as a guidance medium through which the parties can eventually reach agreement as is the case with mediation. Conciliation is popular in Japan[4] because it conforms to the conventions the Japanese are accustomed to using in everyday life. The Japanese tend to prefer settling a dispute through an intermediary. Conciliation between Japanese people need not necessarily be based upon a legal right and would in all probability be settled in a rational manner based upon common sense and morality, although these principles would be less likely to be the guiding ones where a foreign party was involved.

Med/Arb

This is a predominantly US system which is borrowed from labour relation dispute mechanisms and can also take two further forms, 'med-then-arb' and 'arb/med'.

Med/arb has the mediator settling those elements of the dispute that he is able to by mediation, subsequent to which the parties refer the outstanding elements to arbitration. It is usual in med/arb for the initial med/arb agreement to require the mediator to carry on and conduct the arbitration.

In med-then-arb the same processes take place with the exception that two separate neutral mediators are engaged, one for the mediation and one for the residual arbitration.

In arb/med the neutral appointee arbitrates the dispute, makes an award and then mediates the dispute while the award remains in a sealed envelope. If mediation fails the arbitration decision is unsealed, presented to the parties and becomes binding.

To be effective, this system requires the parties to have the fullest confidence in the neutral appointee and they must be prepared to provide whatever confidential disclosures are required by the neutral mediator, although this could have disadvantages in that the possession of too much confidential information might prejudice the mediator's neutrality. A further disadvantage is that the parties may be tempted to withhold information during the mediation knowing that the mediator could later become the arbitrator.

Adjudication

In England and Wales the Housing Grants, Construction and Regeneration Act 1996 (HGCRA) is an example of a statute which gives parties to a contract (in this particular case a construction contract) the right to refer disputes under that contract to adjudication through which the parties can reach a speedy resolution to their dispute in a fair and cost effective manner. The adjudicator may already be named within the body of the contract or, if not, may be agreed upon by the parties to the dispute. If the parties cannot agree on a suitable adjudicator the contract should contain a provision for such appointment to be made by a suitable nominating body[5] (this definition of adjudication should be differentiated from the term as applied to a judgment of the court).

Adjudication is not only a non-consensual process arising from a statutory right such as given under the Housing Grants, Construction and Regeneration Act 1996 in England and Wales. It is often a consensual process and many building contracts and contracts for professional services contain clauses that provide for disputes to be referred to adjudication. Frequently the professional bodies in England and Wales (in particular, in relation to construction contracts, the RIBA, RICS and the Chartered Institute of Arbitrators) are named in such contracts either as the adjudicating or the nominating body.

France is considered founder of the modern system of administrative courts, which manage adjudications, whereas such quasi-judicial adjudication by an agency was abhorrent to the English legal system for a long time, steeped as it was in the Common law. The gradual evolution of administrative adjudication in the US through the executive branch of government sharply contrasts with establishment of the judicial branch. Although administrative adjudication developed most rapidly in the 20th century, it is not of 20th century or even 19th century origin, rather it came into its own in the US in the 18th century. Shortly after the US Constitution became effective in 1789, agencies were established that had adjudicative and rulemaking powers to impose duties on imports into the US. Adjudicators in the US are able and expected to exercise discretion in reaching their decision and they have always had broad latitude to determine the facts in a case. Unlike appellate court decisions, rulings by administrative agencies generally do not have value as precedents.

Adjudication, although popular in most Common law countries, has met with varying support in other countries. In China, for instance, adjudication has not as yet been widely accepted for use, although it has been mentioned in several major construction contracts, particularly those funded by international sources. One of the difficulties in China is that if one party is not in agreement with the adjudicator's decision or simply refuses to agree to implement it, no-one really knows with any certainty how the Chinese courts would approach enforcement, or even whether they would accept jurisdiction over the dispute.

Expert determination

This is most often associated with technical disputes of fact that benefit from an independent expert's knowledge. The expert may be appointed by the parties or, failing their agreement on a suitable person, by one of the appointing bodies such as those already noted in respect of mediation. The expert, once appointed, will be provided with statements from the parties setting out the dispute issues that they require the expert to determine. The expert will disclose each party's submission to the other and the parties will then have the opportunity of issuing formal responses to and questions on the other party's submission.

The expert's determination will be based on the information provided by the parties and will be binding upon the parties. The expert is not usually obliged to give reasons for or with the determination but such a requirement could be written into the terms of engagement should the parties so desire. The Japanese are happy to use expert determination as a means of settling a dispute because it again conforms to the conventions they are accustomed to using in everyday life, similarly to conciliation mentioned earlier.[6]

Early neutral evaluation

Where the parties to a dispute are each confident of their position to the extent that they do not consider mediation or conciliation to be a viable means of resolving it, they may appoint a neutral evaluator, often a practicing or retired lawyer, to assess their respective chances should the dispute go to litigation. The parties submit a joint statement of the dispute initially and follow this with written or oral submissions. The evaluator will then question them on their submissions and will arrive at a non-binding determination of the likely outcome of the disputes under litigation. This is a totally confidential process, less focused on a commercial solution than on a realistic, quick and reasonably authoritative determination of the parties' legal positions.

Mini-trials

Mini-trials are a form of non-binding settlement proceeding developed to resolve disputes between business entities. Mini-trials are an entirely voluntary process and can be written into a contract either as a stand-alone dispute resolution process or as a preliminary attempt to resolve the dispute before resorting to arbitration. The process is particularly suitable for larger, more complex disputes where there are points of law or construction combined with substantial issues of fact but it can be more expensive relative to other ADR techniques, rendering mini-trials unsuitable where the amounts in dispute are small.

Each party to the dispute nominates an executive from its company who is familiar with the dispute and who, together with a neutral chairman usually possessing sufficient expertise in the substantive law at issue to advise the selected executives, form a panel that listens to the facts of the dispute and then tries to negotiate a final settlement. Generally, the parties also agree to staying any pending litigation until the mini-trial is concluded.

The mini-trial is confidential and is preferably scheduled after limited discovery has occurred. The proceedings consist of each side's presentation of a summary of evidence and testimony together with the arguments of their representing attorneys.

At a minimum, mini-trials should succeed in narrowing the gap between the various parties' opinions concerning the issues in dispute, and if settlement is reached they also reduce litigation costs. By involving the parties' decision-makers early on, mini-trials encourage settlement without undue damage to the ongoing business relationships between the disputing parties and the fact that the panel's executive members have a stake in the expeditious resolution of the dispute can only enhance the chances of success. As with other settlement proceedings, mini-trials may not be effective where the parties have substantially different bargaining power or the dispute turns on the credibility of witnesses.

Mini trials are not particularly common in England and Wales, or in fact any other jurisdiction other than the US, although they have found their way into the ADR sections of courses at several English universities.

Dispute Review Boards (DRBs)

These are, like mini-trials, particularly favoured in the US and are gradually becoming used more and more on international contracts in locations other than the US. A DRB[7] is a panel, typically of three people, one of whom is selected by the employer, one by the contractor and the third by the two appointees. DRBs are generally put together at the outset of a contract and work together throughout the contract as needed. They generally meet at site on a regular basis and are kept informed of events as they occur. In that way, once a dispute arises they are in a unique position to focus quickly on the items in dispute with knowledge of the preceding events that they would otherwise have to be taken through in detail.

The DRB is usually run informally without transcripts but the actual structure and modus operandi will typically be determined by the DRB itself. The parties are free to accept or reject the findings of the DRB and may continue their dispute to arbitration or litigation if they so wish. In Hong Kong, one of the first major projects to use a DRB was the new Hong Kong International Airport on Lantau Island and it was so successful that the government will undoubtedly use DRBs again on large projects where the high cost factor will be justified by the enormous savings that can be achieved when a dispute is avoided or dealt with efficiently by the DRB.

ADR generally

In the US there are more than 20 different types of ADR in use within the construction industry and many of these have been successfully used on projects in other countries. The whole idea of ADR is to have a flexible system which suits the particular contract parties, which suits the particular project, which does not conflict[8] with any of the laws governing the contract or its execution and which can contribute to the settlement of disputes within the shortest possible time frame at minimum cost and disruption.

There are several institutions providing different types of ADR services, including:

- Chartered Institute of Arbitrators, UK;
- London Court of International Arbitration, UK;
- Royal Institution of Chartered Surveyors, UK;
- American Arbitration Association, United States of America;
- Hong Kong International Arbitration Centre, Hong Kong;
- Institute of Arbitration and Mediation, Australia;
- International Chamber of Commerce, France;
- Arbitration Foundation of South Africa, South Africa.

ADR should be considered before engaging in the more traditional dispute resolution approaches of arbitration or litigation as long as the particular contract, the dispute that has arisen under it and the rules and regulations in force in the jurisdiction in which the dispute has arisen or will be settled are favourable to such a dispute resolution method.

1. See the footnotes on the first page of the Mediation section of the *LCIA Arbitration-Mediation Handbook*.
2. See the LCIA footnote immediately preceding that referred to in footnote 1 above.
3. The London Court of International Arbitration, the Royal Institution of Chartered Surveyors and the Chartered Institute of Arbitrators are three bodies that provide such a service.
4. Conciliation in Japan is governed by the Civil Conciliation Act 1951 No.222.
5. The Chartered Institute of Arbitrators is such a nominating body and maintains a panel of available adjudicators specialising in different areas of surveying, architecture, building, engineering and law.
6. Settlement would be by the Contract Disputes Resolution Committee established under the Construction Business Act 1949.
7. A 'DRBDC' would be a Dispute Review Board operating during construction whereas a 'DRBAC' would be a Dispute Review Board operating after completion.
8. Conflict with local laws could arise, for instance, if ADR were to be used on a dispute in Saudi Arabia that included a Saudi Arabian party. In Saudi Arabia, commercial disputes are dealt with by the Commercial Circuit of the Grievances Court (Diwan Al-Mazalem), which took this responsibility over from the Committee for Commercial Disputes on 31 December 1987.

Ambiguities, Conflicts and Inconsistencies

There are certain to be instances in the administration of any project, particularly one of an international nature using non-standard hybrid, country or company specific contract agreements, when ambiguities,[1] conflicts and/or inconsistencies are discovered within and between the various documents comprising the contract. At that point the party raising the issue will have to put forward its position and substantiate the reasoning behind its choice of one particular meaning or interpretation.

Hopefully there will be a clause in the contract that defines the order of precedence to be adopted in the event that there is any apparent ambiguity, conflict or inconsistency among the contract documents or any parts thereof. It should include, as appropriate, the Special Conditions of Contract, the General Conditions of Contract, the Schedules of Deviations (if any), the Agreement, Technical Specifications, Drawings, Schedules of Quantities and Prices (possibly Bills of Quantities) and any Appendices attached to the contract documents, the order of precedence possibly being in this order. Such a clause should be thought of as a road map, directing the parties through the various documents so that they can have a reasonable chance of finding which of the words, phrases, clauses or definitions in dispute is the correct one to be used in the interpretation of the contract.

Unfortunately this is not a perfect world and often no such clause is included. There might be a modified version of it or none at all, in which case rules of construction[2] have to be used in order to try to establish from the contract what the intention of the parties was when they entered into the contract.

If the intention is not clear in plain, ordinary words then the parties will have to look deeper. Maybe the particular dispute could be decided by applying the order of precedence of the various contract documents concerned or by adopting the interpretation normally used in the trade or business concerned. If this is not possible the parties must explore all other rules of construction, including expressio unius[3] and contra proferentem.[4] These various steps should result in a positive agreement being reached between the parties, failing which the dispute resolution procedures set down in the contract will inevitably have to be initiated.

The German Civil Code takes matters a step further and actually has a whole section[5] devoted to the use and application of standard business terms with a view to avoiding conflict. In particular:

- Article 305b Precedence of individually negotiated terms

  Individually negotiated terms take precedence over standard business terms.

- Article 305c Surprising and ambiguous clauses

  Provisions in standard business terms that in the circumstances, in particular in view of the outward appearance of the contract, are so unusual that the contractual partner of the user could not be expected to have reckoned with them, do not form part of the contract.

In case of doubt, standard business terms are interpreted against the user.

In addition, under the German Civil Code the above rules cannot be written out of the contract and will apply whatever qualifying conditions are inserted in an attempt to prevent their application.[6]

It is always preferable to write clauses in a clear and concise fashion, avoiding words that can be misconstrued or misinterpreted. It is also advisable to provide a comprehensive list of definitions of words used in the contract to avoid later dispute.[7]

I recall once observing an argument between a team of Japanese colleagues for whom I had recently prepared a Confidentiality Agreement. I ventured across to find out what the major problem was, fearing that I had omitted some important point. The phrase that was causing all the trouble referred to our confidential documents being 'put at the other party's disposal'. My Japanese colleagues were concerned that their confidential information was going straight out with the rubbish, the English to Japanese dictionary they referred to having clearly defined 'disposal' in terms of 'conversion of waste matter'. This was a salutary lesson to learn and I now appreciate more than ever that, although English is my native tongue, the every day application of words with which I am familiar need not be the only interpretation that could be applied to them.

Non-native English speakers would do well to double check any words or phrases they come across which stand out as suspect in their understanding and to remember that when dealing in international contracts it is wise for both native and non-native English speakers to avoid any words which could have a double meaning or which, when translated into a foreign language, could be open to misinterpretation.

Neither should allow themselves to fall into the trap of trying to be 'clever', showing off their proficiency in English, although native English speakers may well find that non-native English speakers have a better grasp of the basics of English grammar than they do. I learnt more about the different verb forms, participles and so on when working for South Koreans than I ever learnt in school.

1. If the ambiguity is obvious it is called 'patent', and if there is a hidden ambiguity it is called 'latent'.
2. See 'Contract – Maxims and Rules of Interpretation' section on p.69.
3. Short for 'expressio unius est exclusio alterius' meaning 'the express mention of one thing rules out other things not mentioned'.
4. See 'Contra Proferentem' section on p.64.
5. *German Civil Code Book 2 Law of Obligations*, Section 2 'Shaping contractual obligations by means of standard business terms'.
6. *German Civil Code* Article 306a – No circumvention: 'The rules in this section apply even if they are circumvented by other arrangements'.
7. See 'Glossary of Terms' in Appendix 1, p.251.

Applicable Laws

It is a fundamental requirement in every contractual situation to establish the law to which the contract is subject at the earliest opportunity. This should be attended to when drafting the contract and should not under any circumstances be delayed until a dispute has arisen between the parties to the contract.

In international trade between companies from different countries their basic choices as regards the applicable law are to either:

- adopt the law of the exporting country – generally this works out to the seller's advantage;
- adopt the law of the importing country – generally this works out to the employer's advantage; or
- adopt the law of a third, independent country – this has advantages of neutrality and can also be useful where the arbitral tribunal defined within the contract is based within that same neutral country.

The parties are generally free to write the terms of the contract in whatever manner they please, and to create obligations one to the other on condition that the intention expressed is bona fide and that the parties do not abuse or contravene rules relating to public order or to laws to which their contractual relationships should adhere (ordre public,[1] lex fori[2] and so on). Their expressed or implied choice of law is a matter of fact, not a matter of law.

There is no universally adopted system to decide upon the applicable law of a contract where the applicable law has not been specifically defined within that contract, however there are a number of conventions and statutes which address this issue.

The Hague Convention deals with matters that can generally be termed international sale of goods whereas the Rome Convention deals with the law applying to contractual obligations. In addition, England and Wales have enacted the Contracts (Applicable Law) Act 1990 which came into force in April 1991. This deals with the choice of law problems in contracts and enabled the ratification of the Rome Convention, which was opened for signature in Rome on 19 June 1980 and was signed on behalf of England and Wales on 7 December 1981. The Contracts (Applicable Law) Act 1990 also gives force to the Luxembourg Convention and the Brussels Protocol. The effect of the latter gives the European Court of Justice jurisdiction to give rulings on the interpretation of the Rome Convention.

For an argument to be accepted to the effect that the contract between the parties evinced an implied choice of applicable law having been made, such implication would have to be unambiguous (under the Hague Convention) or reasonably certain (under the Rome Convention).

The parties may, under the Rome Convention (and the English Act), select the law applicable to the whole of the contract or to a part only of the contract, and may alter their choice of law after the contract has been made. This would give rise to fears of the contract's initial validity being questionable were it not for a proviso which confirms that a contract validly made cannot be rendered invalid by a variation of applicable law, nor can such variation prejudice the rights of third parties.

The Hague Convention proposes the applicable law as being the law of the country where the seller has his business establishment or habitual residence at the time of receipt of the buyer's order and this will govern the contract of sale. Business establishment refers to a permanent and stable business organisation owned by the seller. An in-country agency, a trade-stand at an exhibition or a hotel address would not be sufficient to qualify as a business establishment in this respect. If the seller should have more than one business establishment in the same or different countries, the business establishment receiving the order is the relevant one.

The Rome Convention proposes the applicable law as being the law of the country with which the contract is most closely connected, which could be further defined as where the party who is to effect the performance which is characteristic of the contract has his habitual residence, or, in the case of a body corporate or unincorporate, its central administration, at the time of conclusion of the contract. The Rome Convention can apply to conflicts between two co-existing systems of law within the same State whereas the Hague Convention only applies to international conflicts between jurisdictions of different countries.

When it comes to identifying the applicable law to be applied to an arbitration hearing it must be borne in mind that there are four possible laws involved, namely:

(i) law governing the contract which creates the substantive[3] rights of the parties in respect of which the dispute has arisen;
(ii) law governing the obligation of the parties to submit the dispute to arbitration and to honour the award (the law applying to scope, validity, constitution of tribunal, jurisdiction and so on;
(iii) law governing the conduct of the individual reference;
(iv) law governing the contract which regulates the individual reference to arbitration (whether the parties have been discharged from their obligation to continue with the reference of the particular dispute etc.) may be different from governing law.

Laws in respect of (ii), (iii) and (iv) are usually the same, laws in respect of (ii) and (iv) are rarely different. The governing law of the contract, unless specified otherwise,[4] is the law of the country where the contract is signed and/or executed (relating to constitution of clauses and so on). The governing law of the arbitration is the law of the country where the arbitration is held (dealing with jurisdiction etc.). The procedural law of the arbitration (law of evidence, law of pleading, law of jurisdiction etc.) may be different from the governing law of the arbitration.[5]

In conclusion, the applicable law should always be stated in the contract so that there is no room for doubt. In order to reduce the risk of later disputes it is suggested that a clause be incorporated into the contract defining in no uncertain terms the law that is to apply to the contract and in accordance to which the contract is to be construed. In the US, in the absence of anything specific in the contract, the governing law to be applied to a contract is the law of the state where the contract was made, not where the work is to be carried out.

1. A matter of public policy. This is a term that emphasises the special importance of certain mandatory rules, signaling that they cannot be departed from by anyone and that in principle the courts must apply them automatically.
2. The law of the forum, or the court, in which a case is tried; the law relating to procedures or formalities in force in a given place.
3. Substantive law is defined as 'the basic law of rights and duties (contract law, criminal law, accident law, law of wills etc.), as opposed to procedural law (law of pleading, law of evidence, law of jurisdiction etc.)' (*Oran's Dictionary of the Law*).
4. On large, international projects and on project-financed jobs, the contracts are often noted as being subject to the law of New York, where the financing parties and their legal counsel often are located. Despite §5-1401 of the New York General Obligations Law's requirement that a contractual provision for New York law as governing law be enforced regardless of whether the agreement bears a reasonable relationship to this state, several federal courts have persisted in requiring a sufficient connection between the agreement at issue and New York before upholding a contractual stipulation of New York law and a federal court sitting in diversity in New York still may set aside the parties' choice when such a relationship is lacking.
5. *James Miller* v *Whitworth Street Estates* (1970) AC 583.

Assignment and Novation

Assignment is the transfer of property, rights in property or money or any tangible assets, or any contractual right, to another person or body. The assignment may be brought about by the operation of law or by an action of the party to the original contract who was entitled to those rights now to be assigned, and who now gives the assignee the right to sue upon the contract itself. The general rule is that although contracts may be assigned to another party, the assignor's liabilities under that contract cannot be assigned without the consent of the other party to the original contract and must involve the release of the original debtor. Collins MR is quoted[1] as stating:

> *It is, I think, quite clear that neither at law nor in equity could the burden of a contract be shifted off the shoulders of a contractor onto those of another without the consent of the contractee. A debtor cannot relieve himself of his liability to his creditor by assigning the burden of the obligation to someone else; this can only be brought about by the consent of all three, and involves the release of the original debtor.*

An exception to this general rule would be where the 'liability' is annexed to the benefit, making it an indivisible part of the rights assigned. In the case of *Aspeden v Seddon*[2] the right to extract minerals was subject to a duty to pay compensation should the land subside. This liability was found to have been passed to the assignee of the mining rights as it was a condition for the enjoyment of the benefit.

Assignments may be 'absolute' whereby the assignor transfers all of its rights under the contract without restriction or qualification, or 'conditional' whereby the commencement or cessation of the assignment is dependent upon the occurrence of some future but uncertain event.

An assignment may be by way of a charge only, whereby the right is not actually transferred, or may be of part only of the rights under the contract. Absolute assignments may be recognised as statutory assignments (under which the assignee can sue in his own name without joining the assignor) whereas conditional, partial and charge only assignments may not. These latter cases may, however, be recognised as equitable assignments even if not in writing and not accompanied by consideration. Under an equitable assignment, unlike under a statutory assignment, the assignee does not acquire a legal title to the chose in action[3] assigned and the assignee cannot sue upon the contract alone but has to join with the assignor. The assignor in turn cannot sue alone once the assignment has been made but must join the assignee in its action.

When an employer is assigning his interest in a contract, the contractor should be particularly aware of:

- conditional assignment whereby the contractor would be linked with both the assignor and the assignee, particularly in respect of performance guarantees, payments, final acceptance and so on – the contractor would need to be given

a clear explanation of responsibilities and obligations of assignor and assignee to avoid disputes at a later date;
- partial assignment whereby the employer only assigns its rights and obligations in respect of a part of the original agreement.

It is perfectly acceptable to include an express provision in a contract to the effect that the rights under the contract shall not be assigned. In such a case any purported assignment will not only be a breach of contract but also of no effect and will not confer to the assignee any rights against the debtor.[4] There are circumstances, however, where such an assignment, although expressly excluded under the contract, may yet be binding upon the assignor and assignee.[5]

Assignment is not a concept that applies only under Common law but is also applied in Civil Code systems. The main features of assignment under the French Civil Code,[6] for example, are basically the same as those under English law but without the complexities of the distinction between equitable and statutory assignment. The Code places assignment under the heading of 'The Assignment of Claims and Other Incorporeal Rights'.[7]

The situation whereby liabilities are transferred to a third party with the consent of the other party to the original contract is called 'novation' and is evinced by a tripartite agreement between the original parties to the contract and the new party. Agreement to novate may be express or implied (for example, where ownership of a business is transferred and the existing customers continue to trade with the new owners they are deemed to have accepted the novation by implication).

In novation, a new contract is substituted for the old one. The parties to the new contract may be the existing parties (for instance, where an agreement in writing is added into a deed at a later date) or different parties (for example, where a new party takes the place of the original debtor or creditor whose rights and liabilities then cease to exist). If, for example, one party to the contract owes the other party a specific sum and that other party in turn owes a similar sum to a third party outside the contract, the parties can agree to enter into a tripartite contract whereby the first party agrees to pay the debt to the third party thereby extinguishing the second party's liability.[8] The new contract must make it clear that the debt owed by the second party to the first party is extinguished otherwise the new contract would fail for lack of consideration (a novation agreement will only be binding if supported by consideration, except in the case of a 'Deed of Novation', no consideration being necessary for agreements executed by deed).

It is common in construction design and build contracts for the employer to require the contractor to accept a novation of the original architect or engineer's appointment and to take, therefore, full responsibility for all the design that has been carried out up to the date of novation. This is full of risks for the contractor, especially on the more complex projects and where the tender period is too short for it to be able to carry out detailed checks on the design that it is taking responsibility for. The contractor has a lot more at stake than the employer did when it signed the original agreement with the designer, including exposure for the extra costs incurred in remedying the defective work attributable to defective design and for its resultant loss of profit. In a Scottish case,[9] the novation agreement stated that:

> *The liability of the Consultant under the appointment, whether accruing before or after the date of this novation shall be to the Contractor and the Consultant agrees to perform the appointment and to be bound by the terms of the appointment in all respects as if the Contractor had always been a party to the appointment in place of the Employer.*

The employer had suffered no loss as a result of the design fault and had expressed no dissatisfaction with the consultant engineer's services. It was ruled that losses suffered by the contractor as a result of the engineer's breach of its appointment prior to the date of novation would not have been suffered by the employer had there been no novation, as the contract was a fixed price contract and passed the design responsibility on to the contractor. The novation agreement gave the contractor no better rights than the employer had originally. (i.e. the contractor could not recoup losses suffered due to the engineer's breach).

As noted, this was a case under Scottish law and its application under other laws, including laws of England and Wales, may be subject to dispute. It would be wise to cover the contractor's exposure to the possibility of such a risk by ensuring that there is an express provision in the novation agreement to enable it to recover the full extent of losses which it might incur, not merely those which the employer would have recovered under the original agreement, which might well have been more limited.

Where a novation agreement is being considered in respect of an engineering or design consultancy agreement, the contractor would be wise to study the agreement to be novated in some detail and to take whatever steps are available in order to minimise its risk. Having probably been in the design 'driving seat' for some time prior to the intended novation, the consultant will not be likely to be cooperative in adding further conditions to its appointment on its novation to a contractor and the contractor must take this into account when preparing its tender risk analysis. Ideally employers should also take this potential issue into account when preparing their initial consultancy appointment agreements in order to avoid the burden of excessive risk allowances being built into the tenders and in order to avoid future conflicts arising during the contract execution stage.

Under the French Civil Code novation is noted as brought about in three ways:[10]

- where a debtor contracts towards its creditor a new debt which is substituted for the old one, which is extinguished;
- where a new debtor is substituted for the old one who is discharged by the creditor;

- where, by the effect of a new undertaking, a new creditor is substituted for the old one, towards whom the debtor is discharged.

Whereas the above is similar in principle to novation under English law, the application differs from that in England and Wales in that under the French Civil Code novation may be brought about by the substitution of a new debtor without the assistance of the first debtor.[11] Also under the French Civil Code 'Novation may not be presumed; the wish to bring it about must clearly result from the instrument'.[12] This differs from the idea of novation by 'implication' or 'deemed acceptance' accepted under English law.

1. *Tolhurst* v *Associated Portland Cement Manufacturers (1900) Ltd (1902)* 2 KB 660 at p.668.
2. *Aspeden* v *Seddon* (No. 2) (1876).
3. A 'chose in action' is a right of action that can be enforced in a court of law. Assignment of a contract is deemed to be an assignment of a 'chose in action'.
4. *Linden Gardens Trust Ltd* v *Lenesta Sludge Disposals Ltd* (1994) 1 AC 85.
5. *Don King Productions Inc.* v *Warrant* (1999) 2 All ER 218.
6. The French Civil Code equivalent of Assignment is called 'cessation de créance'.
7. French Civil Code, Articles 1689 to 1701.
8. *Scarf* v *Jardine* (1882) 7 App. Cas. 345.
9. *Blyth & Blyth* v *Carillion Construction Co. Ltd* (18 April 2001).
10. French Civil Code, Article 1271.
11. *ibid.*, Article 1274.
12. *ibid.*, Article 1273

Bonds and Guarantees

A bond in construction contract terms is basically the same as a guarantee and for the purposes of this section the two words will be treated as synonymous. It is basically a three party contract whereby the surety company (usually a financial institution) or bondsman provides an irrevocable guarantee to the employer that the contractor will faithfully and dutifully complete its obligations under its contract with the employer. The three parties are the employer, the contractor and the surety. Unlike an insurance policy, which is based upon the premise that there will be a loss and which charges a premium designed to cover that loss, a bond is an extension of credit on the basis that the legal obligation bonded will be fulfilled and that there will be no loss. Bond premiums are designed to cover the surety's underwriting expenses only.

Bondsmen do not generally have much risk as after they have responded to the call upon the bond made by the employer they have immediate recourse to the contractor. Unless the contractor has become insolvent the bondsman will recover the monies it has paid out. It has been known for a contractor to be successful in recovering the amount of the bond proceeds where an unfair call was made.[1] There are also precedents that enable the recovery of the excess of any call over the actual loss suffered by the employer.[2]

There are two generic types of bond, namely 'default bonds' (performance bonds) and 'on-demand' bonds (which includes the hybrid 'conditional on-demand' bond). Within these generic types there are five principal bonds most frequently used in construction contracts, and these are:

- Tender (or Bid) bonds;
- Advance Payment bonds;
- Performance bonds;
- Retention bonds;
- Warranty (or Maintenance) bonds.

(In the US there is also a requirement for bonds from a main contractor guaranteeing payments to its sub-contractors and suppliers.)

A ***tender (bid) bond*** guarantees that the bidder will enter into a contract to carry out the works at the bid price. If the tenderer withdraws its bid at any time before bid opening (or at any such similar time as might be specified within the contract) then the security shall be returned to the bidder. If, however, the tenderer withdraws its tender after the bids have been opened and if the reason for this is other than because a non-judgmental mistake was made in the bid, the bid security will be forfeited.

An ***advance payment bond*** guarantees the reimbursement of the advance payment should this not happen in the normal course of events (such as a deduction from interim monthly or milestone payments due to the contractor).

A ***performance bond*** guarantees performance by the contractor and typically protects against insolvency, completion, attainment of specified achievable levels of plant output and so on.

A ***retention bond*** is sometimes given to the employer in place of the employer deducting a percentage retention from each interim and milestone payment due to the contractor, not to be reimbursed in full until satisfactory achievement of completion of the defects liability period. This may be issued on a reducing basis, commonly being reduced by 50 per cent upon achievement of substantial completion.

A ***warranty bond*** guarantees the contractor's performance during the warranty period, putting right any defects or warranty issues that arise. It is intended to protect against design defects and/or failures in workmanship or materials/equipment. Warranty bonds are limited to a fixed period of time, usually the same as the warranty defects/maintenance period established under the contract.

Default (conditional) bonds put an obligation on the issuer to pay provided that the employer can provide sufficient proof both that the contractor has defaulted on a term of the contract and that the employer has suffered damages as a result.

On-demand (unconditional) bonds put an obligation on the issuer to pay, as the name suggests, on demand without the need for proof of default and even without any default or breach having occurred. The principle is 'pay now, argue later', and the only resistance to such a bond is for the contractor to prove fraud.

Bond language is often very complicated and confusing, which is totally unnecessary as a bond merely has to say, for instance, that it guarantees to the employer performance by the contractor under a specific contract up to a maximum (quoted) liability for a certain period of time. Many words are synonymous and are used in different combinations in different countries: bond is synonymous with guarantee; guarantor with surety, obligor, (issuing) bank and bondsman; principal creditor with obligee; beneficiary with employer; principal debtor with contractor, sub-contractor or supplier as the case may be; performance with default; and unconditional with on-demand.

The wording used in defining the exact type of bond to be provided is extremely important and it is wise to try to keep it as simple as possible whilst still conveying clearly the intentions of the parties. The following check list is useful in this respect:

- the contractor should try to ensure that it has the opportunity to defend itself against a call;[3]
- state clearly whether the bond is on demand or conditional, do not leave this for the courts to determine – remember that an on demand bond will involve paying now and arguing later;
- use uncomplicated language consistent with the contract wording to minimise confusion;

- bonds must be in writing to be enforceable;
- clear drafting of the conditions (conditional bonds) is crucial;
- establish and state clearly the currency unit and the currency value of the bond;
- establish and state clearly the duration of the bond.

It is usual for the contractor to bear the cost of obtaining bonds and for any other expenses in setting them up successfully.

1. *ENS Limited* v *Derwent* (1998) 62 ConLR 141.
2. *Cargill* v *Bangladesh Sugar and Food* (1998) 2 All ER 406.
3. It is perfectly acceptable for a clause to be included which:
   - requires the employer to give the contractor notice within, say, 20 days of its intention to draw down on security funds;
   - requires the contractor to, within a further period of, say, 20 days, either accept the employer's decision under its notice by issuing a notice of no-contest, or to contest the notice, in the latter case setting out its details for doing so;
   - requires the employer to either accept the contractor's contest and withdraw the notice or to give notice of a dispute (in which case the dispute resolution procedures set out in the contract would be brought into action);
   - confers the right upon the employer to deem acceptance of its initial notice if the contractor fails to respond to it within the allotted period;
   - stipulates that draw down can only be made upon either a notice of no contest or deemed acceptance.

Change Order Management

Effective management of variations and change orders plays a critical part in project and contract management from perspectives of both time and money. Effective variation and change order management starts during the bid process and continues until the project works are finally completed. It has applications in most types of contract whether they are EPC, traditional lump sum with or without bills of quantities, cost-plus or guaranteed maximum contracts. For the purpose of this section, variation orders and change orders will each be referred to as 'change orders'.

The first step in change order management starts when the bid documents are reviewed and any aspects of them which the contractor considers to be either ambiguous, unclear or un-constructible are reported to the employer, preferably well in advance of the tender due date. This will give the employer time to make a decision and, if necessary, to inform the other bidders of its decision. Through resolving these issues pre-bid, the contractor and employer may well have avoided the need for a change order to be issued during the currency of the contract, and possibly have also avoided a subsequent dispute.

The bid documents must be thoroughly checked to ensure that there are not only clauses which give the employer the opportunity and the means to issue change orders as the project proceeds should the need arise (such as changes in design, quantity, quality, levels, sequencing of works and so on), but also that give the contractor the opportunity and means for proposing a change order if it considers one to be either beneficial to the employer and/or the project, or for asserting that in its opinion an instruction or direction of the employer or a particular event has given rise to the need for a change order to be issued.

I recall a particular contract signed by a Korean contractor in Saudi Arabia which included a clause stating that the contractor would 'carry out any and all variations to the contract works as may be instructed in writing by the Employer at no additional cost and within the time allotted in the contract for the completion of the Works'. The unfortunate contractor had failed to understand the full implications of this clause when it signed the contract and had no way out of it. It would probably have stood up in courts in the US, England or continental Europe (depending upon the full extent of the additional works ordered) despite being clearly unreasonable, since it was in no way ambiguous, unclear or written in small print. Thorough checking of the bid documents by a person familiar with contracts would have enabled the contractor to draw this to the employer's attention before it submitted its bid.

The contract clauses should define exactly what the scope is for change orders and should further define a clear process for the initiation, processing, instruction and application of change orders. Both the employer and the contractor should have established procedures in place to ensure that there is no slip up in any of these respects.

There are basically six steps in change order management that must be considered in each and every case if management is to be truly effective. These are:

- Step 1 Employer issues request for change order proposal to contractor in a timely manner.
- Step 2 Contractor responds promptly to employer's request.
- Step 3 Employer reviews contractor's response promptly.
- Step 4 Employer issues change order.
- Step 5 Contractor promptly performs the work covered by the change order.
- Step 6 Employer pays the contractor promptly upon receipt of an invoice in respect of change order works.

These six steps will now be dealt with in more detail.

Step 1. Employer issues Request for Proposal (RFP) to contractor in a timely manner

It is extremely important for the employer or the employer's representative to issue change order proposals at the earliest opportunity. The later that a change is proposed, the more the effect it has on the contract schedule, the more difficult it is to perform and the more costly it is likely to be. Whether the change order proposal is wholly initiated by the employer, or whether it results from the contractor having identified a need for a change and having so informed the employer, the change order proposal should be formalised as soon as possible.[1] Apart from making a direct request for a change order, it is perfectly in order for the contractor to signify on an RFI (Request for Information) the possible cost implications of the matters covered by the RFI. This is good contract management, dealing with the issues in a business-like fashion as soon as it is possible to do so, leaving the employer to decide whether to proceed with the additional cost works as a change order or to revise its requirements to avoid any change implications.

Step 2. Contractor responds promptly to employer's request

It is in both parties' best interests for the contractor to respond to the employer's RFP promptly upon its receipt. The contractor should have procedures in place to deal with each RFP promptly and effectively. Where a sub-contractor is involved the contractor must have procedures whereby the relevant parts of the RFP are forwarded to the sub-contractor immediately. As soon as the feasibility of the actual change proposal has been established, the contractor's project team should start working on an estimation of cost. It might be that the contractor is unable to give a firm proposal at that particular time, in which case a provisional assessment will have to suffice. The employer and the contractor would both benefit from having agreed on RFI and RFP procedures at the outset of the contract, so that the employer knows what information the contractor needs in order to estimate the time and cost impacts of the RFP and the contractor knows how the employer would like its response presented.

Step 3. Employer reviews contractor's response promptly

The employer should have a similar procedure for receiving and checking the contractor's response to an RFP and should ensure that this procedure is implemented promptly upon its receipt. It is preferable for the employer to call the contractor to a meeting if there is anything in the contractor's proposal which the employer cannot understand or which it considers to be in error. This is less time-consuming than returning the response for correction and amendment and is a far more business-like approach, which is more likely to encourage the contractor's cooperation and speed up the whole change order procedure.

Step 4. Employer issues change order

When the employer is satisfied that the response to the RFP is in order and that the contractor's proposals in terms of technical content, time and cost are realistic and reasonable it should issue the change order. In practice there are many situations where the issuance of a change order is a time-consuming exercise sometimes needing the verification of a party or parties other than the party carrying out the review, such as the lenders or the lender's engineer. In such cases some form of work authorisation should be issued confirming that a change order has been agreed and that the contractor is authorised to commence the change order work while the formalities of issuance are being attended to.

This authorisation document should also confirm and, as necessary, set out the scope and the agreed cost of the work covered by the authorisation together with the terms for payment that will operate until the change order has been officially issued. On occasions the employer and the contractor are unable to reach a timely agreement on the proposals but it is imperative to the timely continuance of the contract that the change order works are proceeded with without delay. Most contracts make provision for this situation and usually require the contractor to keep time, material, plant, equipment and other relevant records concerning the change order works to help in resolving the cost and time issues should the parties fail to agree. These records are usually kept on a 'daywork'[2] record form signed on a daily basis by the contractor and the employer's representative as being true records.

Step 5. Contractor promptly performs the work covered by the change order

The contractor should act promptly in the performance of the change order work and should execute it as soon as possible after the issuance of the order. The contractor will in all probability, if the change order works are significant, have included a schedule revision as part of its proposal. Once the change order has been issued the contractor should check this revised schedule to ensure that it is still valid and should re-issue the project schedule incorporating the change order works. It is advisable to identify the change order work activities clearly on the revised project schedule so that the employer can see clearly the effect they have had on other activities.

Step 6. Employer pays the contractor promptly upon receipt of an invoice in respect of change order works

The employer should pay the contractor for the change order work just as he would pay for any other work, that is to say in a manner consistent with the payment regime agreed upon and active for the project as a whole. The employer should budget for change orders so that it is able to fund them when they occur. The contractor should not have to suffer from the employer's cash flow problems due to poor budgetary control by the employer.

Pricing change order proposals

When pricing the change order proposal the contractor must take account of all the factors affected by the work it covers. If it is merely a material substitution there might be no obvious change in labour involvement but on closer examination the new material might in fact require, for example, a higher labour, plant or equipment contribution than the original material required, either at the storage, handling, preparation, installation or finishing stages, and the contractor must be able to substantiate this if it is the case. Manufacturers' data will normally provide information which can be used in this respect and the contractor might well have its own records where it has used both the original and the substitute materials on previous jobs and which can be put forward as evidence of the relevant labour, plant and equipment requirements of each.

Where there is additional labour content it is safe to say that there will invariably be a time consideration, although the additional time to be added to the contract period might not be the same as that expended on the change order work. The change order work would have to be plotted on the critical path network diagram in order to see what effect it actually had on other works before a definitive statement could be made in this respect.

The contractor will want to add percentage mark-ups to the direct and indirect costs incurred to cover its overheads (head office and site office) and profit. In the event that the contract has not defined what percentages are to be added to change order proposals, the contractor might have to prove, either from company financial records or possibly the tender rate build up, what the appropriate percentages are. Where work is similar to contract work and where there is a bill of quantities or schedule of rates included in the contract documents, there will hopefully be rates in those documents that can be either used directly or interpolated to suit the new items. It is not unusual to find a contract rate that is clearly in error and a separate section of this book addresses that issue.[3]

If there is a dispute about any aspect of the change order proposal, either in respect of a rate build up or in respect of the proposal itself (for example, the contractor might consider it outside his scope of work) then the contractor, who is most probably obliged under the contract to proceed with the works while a dispute is resolved, should keep accurate records of all time, material, plant and equipment time and costs expended carrying out the change order proposal works. These should be signed by the employer's representative as a true record and can be used at a later date as a basis for agreement should none have been reached in the normal course of events.

1. See typical format for 'Change Order Proposal' in Appendix 8, p.299.
2. See 'Dayworks' section on p.81 and typical 'Daywork Record Sheet' in Appendix 5 on p.291.
3. See 'Tender Rate Errors' section on p.235.

Claim Prompt List

The following prompt list of claim terminology has been found to be an indispensable memory jogger as events have unfolded on the many large-scale international projects that I have been involved in. It has proved itself invaluable in the effective claim management of a variety of different types of construction project in a variety of countries in the Middle East, Far East, US and Europe and its usefulness is by no means restricted to one profession, one branch of the industry or to one locale and is of equal benefit to representatives of the employer and the contractor. The topics have only been touched on briefly in this section; more in-depth explanations of many of these topics are included in other sections of this book.

It is well worth all the members of the project team considering the claim prompt list items as triggers and referring to them frequently in order to be reminded of the main issues that have to be monitored and taken into consideration on a daily basis on a busy construction project. The topics cover not only the claim events themselves but also items which may be included in any claim formulated, and are relevant not only to the employer's and the contractor's professional management teams but also to the site and trade supervisors, programmers, material procurement staff and so on, and each would benefit from a copy being posted on their office wall for reference. When claim triggers are activated the project team should immediately put into operation the claim management procedures and strategies (whether from a claim or claim defence position) which they would have established at the outset of the contract, including serving the appropriate notices and maintaining contemporary records relating to the events concerned.

Acceleration

- Acceleration can occur where the contractor is left with no option but to revise his resource usage to try to complete the project works within the original contract period.
- The contractor will incur additional costs including additional overtime costs, loss of efficiency, cost of introducing additional labour, plant and equipment to the site and the 'ripple'[1] effect that the acceleration has on other trades and other sections of the works.
- The acceleration could either be 'directed' (i.e. ordered directly by the employer) or 'constructive' (where the employer compels the contractor to complete on the original contract completion date and refuses to grant a reasonable request for an extension of time).[2]
- Directed acceleration may occur where the employer uses a specific contract clause to direct acceleration as a means of achieving the date for contract completion set out in the contract.
- The contractor may be entitled to additional payment and to extension of time for that portion of delay not remedied by acceleration where it would otherwise have been so entitled under the contract in respect of that delay had no acceleration been directed.

Additional equipment

- Equipment brought to site in claim situations will have its total cost (including the cost of mobilisation) allocated to the claim; any equipment already on site will only have its operational time charged to the claim.

Additional work

- Normally additional to BQ quantities – will be adjusted at, or pro-rata to, BQ rates where appropriate and where reasonable to do so (see also 'Extra Work' in this section).

Ambiguous clauses[3]

- These may have two ostensibly reasonable interpretations. Unless the contract has specific provisions for determining the correct application, the contractor's interpretation (or sub-contractor's in a sub-contract situation), if reasonable on its own account, should prevail 'contra proferentem'[4] since it had no hand in the preparation of the document and hence no opportunity to resolve the ambiguity at the drafting stage.

Bonding cost

- An additional fee for the extra period or the extra value of the project would be recoverable since this would be additional to that originally budgeted for.

Burden of proof

- The burden of proof is on the party presenting a claim to the other party.

Concurrent delay[5]

- The contractor has to show that its delay only affected float on the critical path network (CPN) whereas the employer's delay was critical to the critical path itself.
- Where both parties are partially to blame neither can claim damages from the other for the 'overlap' portion of the total delay.

Consequential damage

- This may be recoverable – it must have been reasonably foreseeable or within the contemplation of the parties at the time of contracting for it to be so.

Consistency

- Contractors should review sub-contractor claims to ensure consistency with their own.
- Employers should look for inconsistencies in contractor's claims as signals that all is not right with the claim as presented.

Constructive variations[6]

- Where these have the effect of changing the contractor's obligations or responsibilities under the contract they must be responded to by the contractor with a claim for recovery of costs/time in a timely manner or else the contractor's right to claim could well be lost.

Consumables

- Recovery of the cost of consumables (i.e. those materials consumed during the construction process but not incorporated into the project) can be claimed where they are affected by a claim event.

Damages

- Damages recoverable on a breach of contract submission must have been foreseeable at the time of the breach and must have been within the contemplation of the parties at the time they made the contract as the probable result of such a breach.

Depreciation costs

- Those on company-owned equipment may properly be included in a delay claim.
- Not so on hired equipment.

Direct costs

- Direct costs refer to items incorporated into the project, including labour, materials and equipment necessary to carry out the contract works.

Direct labour

- This is labour directly associated with the work product – it includes basic wages, bonuses, fringe benefits (holidays, transport, housing, meals and so on) and social welfare taxes.

Direct material costs

- These relate to raw materials, purchased parts and all works, including subcontracted works, required in order to manufacture, assemble, construct, erect and complete the works, including wastage, overruns, spoilage and defective works when computing the total costs.

Efficiency reductions

- These can result from the introduction of additional crews into already occupied working areas, and are noticeable by workers falling over each other or having to wait for one crew to finish in an area before they can have access to that particular portion of the works.

- Efficiency reductions can also result from moving crews from one job with which they are familiar to another with which they are not familiar and maybe even then back again – learning curves are experienced at each point of change.

Equipment costs

- These include the costs of all machinery and equipment together with necessary supplies for upkeep and maintenance.
- They also include tools and apparatus necessary for the proper construction of the work.

Escalation costs

- These are additional costs incurred when a delay event forces performance at a higher wage rate, higher material cost or higher plant and equipment cost than planned in order to achieve completion within a reasonable time frame considering the nature of the delaying event.
- They are not to be confused with normal increased costs due to market forces and which may be either claimable or non-claimable depending upon the precise wording of the contract.

Extra works *(as opposed to 'additional work' noted above)*

- These are works not previously contemplated so no BQ items or rates are applicable. The employer/engineer will either have to agree new rates with the contractor to be applied on a re-measured or lump sum basis, or will have to agree to pay on a daywork basis.

Finance charges

- Unless expressly precluded by terms of the contract, these are recoverable in relation to, amongst other headings, delayed payments, the filed claim while it remains unsettled and the cost of capital supplied internally and/or externally, calculated in the normal manner usually adopted by banks.

Home office overheads

- These are best calculated by dividing total office overheads by total company direct costs to determine a rate expressed as a percentage, which is then applied to the claim total.[7]
- Extra time will increase overhead costs and these can rightly be included in the overall claim as a reimbursable cost.

Indirect costs

- Include all those costs not directed to one specific site objective but having multiple objectives (such as supervision of whole or parts of specific sections of the site).

Insurance costs

- Additional premiums necessary for the extra period or the extra value of the project, and therefore not contemplated in the premium originally budgeted for, are recoverable.

Jobsite overheads

- These are usually expressed as a percentage of the direct costs.

Loss of efficiency

- This is calculated by comparing normal productivity to actual productivity as affected by the claim event despite the contractor's mitigation efforts.
- It can be a comparison of one portion of the site unaffected by the claim situation against a portion of the site that has been so affected.
- If this is not possible, compare the affected portion with similar work on a previous project or with data from the tender build-up.

Mitigation[8]

- The contractor must show that it has tried to achieve mitigation of losses/ damages.
- Failure to do so might prejudice a contractor's chances of reaching a fair settlement of its claims either by the employer or by the arbitrator, both of whom would look more kindly upon a contractor who had tried its best to mitigate the losses and damages resulting from the claim event.

Notices[9]

- Notices must be served in a timely manner as required by the contract to the address provided in the contract and must include all the data and information required by the contract.

Out of sequence

- Out of sequence working because of disruption of the contractor's programme by the actions of the employer can lead to delays and can create efficiency reductions, with the resulting costs/damages being claimable.

Overall material costs

- Material costs used in claim calculations must make due allowances, where appropriate and relevant, for trade and cash discount, refunds, rebates together with scrap value of sold items.
- They should also include a percentage mark-up to account for overhead costs.

Overhead costs

- These can be claimed in every case and are defined as the costs of management and administration of the site/company both as a whole and in particular in relation to the subject contract.
- Overheads include managers (not field supervisors – these would be classed as direct or indirect costs depending upon whether full time on the one project or supervising several).

Overtime

- Overtime in excess of that reasonably anticipated and budgeted for creates inefficiency.
- Efficiency can be reduced by between 25–45 per cent after working a 50-hour week over a ten-week period.
- The longer the hours, and the period over which longer working hours are imposed, the greater the inefficiency.

Partial disruption of work

- This causes reduction in the efficiency of carrying out repetitive actions, thus necessitating further learning curve adjustments, increasing time and costs.

Presumption

- Presumption is that actual costs are reasonable costs.
- Except in some 'quantum meruit'[10] applications in which case it might be considered more equitable to use current market rates rather than actual costs.

Profit margin

- Profit margin mark-ups to be recovered on contractor's additional costs are not necessarily those included in its original bid, but should be reasonable and should be capable of justification.
- Profit can also be added to sub-contractors' claims where these are being incorporated in the main contractor's claim.

Proving entitlement

- This is a condition precedent to any recovery. The contractor must be able to:
  - prove that the contract was materially breached or that it is entitled to an adjustment under the terms of the contract, stating the clause under which the claim is made;
  - establish the amount of additional time and/or costs incurred;
  - prove a cause and effect relationship between acts and omissions of the employer (cause) and additional time/costs (effect).

Recognising claims

- This depends to a significant extent upon having in-depth familiarity with all of the contract clauses, the drawings, specifications, programmes and other contract documentation and with the day-to-day happenings on site.
- There is no substitute for such knowledge when it comes to identifying or disputing a claim situation.

Records

- Records must be kept in an organised manner, including chronological correspondence files, memo files, minutes of meetings, programmes and revisions, site and office diaries, payroll records, test and inspection reports, quality control records, work procedures and so on.

Rejection

- An employer's rejection of documents submitted for approval, under the contract must be issued within the time for review established by the contract or, if none, then within a reasonable time.
- Rejection of a contractor's claim by the employer should be responded to immediately, by the contractor, with a formal notice at the very least preserving its right to contest the rejection at a later date.

Restricted access

- This reduces the amount of work that can be performed in a given time creating unabsorbed overheads and delay, both of which are claimable.

Ripple effects

- These can have serious implications on events a long way down the activity chain.
- It is advisable to plot ripple effects on a programme intended specifically to highlight them in support of a subsequent claim application.[11]

Sufficiency of resources

- This is the contractor's responsibility; however, events brought about by the actions of the employer can have adverse effects on the progress of work, giving the impression that the contractor has insufficient resources on site.
- This situation should be monitored by the contractor, and it should maintain contemporary records and serve notices upon the employer as deemed appropriate and in accordance with the requirements of the contract.
- The employer should also monitor the situation and bring to the contractor's attention incidents of adverse effects being felt due to the contractor's perceived inadequacy of resources, thus protecting the employer from subsequent claims directed at putting the responsibility for delay unfairly onto the employer.

Work rescheduling

- If as a consequence of an employer action resulting in certain works having to be carried out during a different season from that anticipated or at night time instead of day time or in a different sequence from that anticipated by the contractor, work rescheduling will inevitably result in working procedures changing and, subsequently, additional costs to the contractor.
- An example would be if sub-structure work were to be reprogrammed from the dry season to the rainy season. This would involve, for example, considerably more site work such as dewatering and pumping, reduced efficiency due to the stop/start nature of work during inclement weather conditions, etc. and would result in an increase to both the contractor's costs and the time needed to complete the works.
- The further projects are situated away from the equator the more significant is the change from working in summer to winter. The effects of such a change would be felt for instance by the contractor having to cope with a shorter working day as daylight hours diminish in addition to having to cope with significantly changed temperatures and rainfall conditions.
- In the Middle East a change from winter to summer working would increase the costs of, for example, concrete curing where the inclusion of additives, watering, ice etc. would be requirements that would not otherwise have been called for, or at least not in the same quantities.

1 'Ripple' effect is the effect that disruption in one trade or one section has on other trades or other sections of work (for instance, if blockwork is delayed it has a ripple effect on following trades such as joinery, electrical work, plastering, decoration etc.).

2. See 'Acceleration – Directed and Constructive' section on p.7.

3. See 'Ambiguities, Conflicts and Inconsistencies' section on p.19.

4. See 'Contra Proferentem' section on p.64.

5. See 'Concurrent Delay' section on p.54; In addition, for a concise and easily readable work on this subject refer to John Marrin QC, 'Concurrent Delay', Society of Construction Law (February 2002).

6. 'Constructive Variation' claims are frowned upon in the England and Wales where it is felt that there are sufficient contractual and legal avenues open to a contractor who believes that an employer's directive should be treated as an official variation without recourse to such a claim. In the US and an increasing number of other international jurisdictions, however, it is an accepted claim heading.

7. See the most commonly accepted and used formula applications detailed in the footnote on p.50 in the 'Claim Review and Preparation' section.

8. See 'Mitigation' section on p.174.

9. See 'Notices' section on p.186.

10. See 'Quantum Meruit' section on p.197.

11. See footnote 1 above.

Claim Review and Preparation

It is not the intention to write a definitive thesis on claim identification, initiation, preparation, submission and negotiation in this section since that is a subject on its own and would take volumes of dedicated writing to cover. This section follows on from the previous one, which proposed using a 'Claim Prompt List' to maintain awareness of claims management at all times, and will simply guide the reader through the processes necessary when researching and preparing a claim.

It is assumed that this process is being carried out either some time into the contract or at the end of the contract. It has been stated in other parts of this book that it is preferable to start preparing claims registers and collating all the relevant documents, data and facts from day one of the project, even from the date of receipt of the tender enquiry, and that recommendation is endorsed once more here. If this recommendation is followed, similar procedures to those described here would be adopted and these would hopefully be written into the project procedure manuals to be carried out contemporaneously, not retrospectively as is the case here.

This section has been written from a contractor's perspective, however the steps described will work equally well whether they are being followed by a contractor or an employer and whether for the preparation or defence of a claim, possibly with some minor deviations to reflect the different user and the different nature of the operation (for instance, Step 3 would be more difficult for an employer to develop; Step 5 would relate to the preparation of a 'claim defence document' rather than a 'claim document').

The end result of the processes to be described in this section will be a report on the claim, which will be the document to be submitted to the other party. This will be submitted with the initial intention of negotiating a settlement between the parties. This will not always be the way things work out and it may well be that an agreement cannot be reached and, under the terms of the contract, the dispute has to be referred to arbitration. This will involve lawyers, attorneys, counsel, barristers (different terminologies are used to describe these professionals in different jurisdictions) and technical experts reviewing the facts behind the dispute and preparing their respective cases for submission to the arbitration panel.

Much of this work will duplicate that already carried out in order to prepare the original claim document upon which the failed negotiations were based. It is suggested that the initial claim document should, from the outset, be prepared as if it were to be used in arbitration,[1] thus reducing the workload, and therefore the time, of the arbitration itself. In addition, if all the documents referred to in the claim research are properly tagged, classified and recorded, this will help not only in discovery and evidence but also in enabling all involved in settling the dispute to put their hand on all the relevant facts with ease. This can only help the contractor's case at the end of the day.

Claim specialists each have their own particular way of working but the following approach will, at least in general terms, be familiar to many of them. Taking a systematic approach such as this will develop the means to build up a picture of the claim event visibly supported by records that have been examined, documented and retained in easily accessible databases. The claim justification data prepared from these records, which could be in the form of as-built programmes, bar charts, CPN[2] schedules, financial analyses and so on, will assist the defendants, arbitrators or whoever in their efforts to understand and decide on the cause, effect and quantum of the claim. The party who has the best records, and who is therefore able to justify his claim in this way, will stand the best chance of success, all other things being equal.

Step 1 – Research of claim issues

- Review project documents, including:
  - contract agreements and other documents together with appendices, addenda, specifications, bills of quantities, agreed deviations etc;
  - correspondence between parties to contract, sub-contractors, suppliers, authorities etc. (including letters, faxes, e-mails, memos, transmittals, etc;
  - minutes of meetings;
  - design and as-built drawings, including all submittals, revisions, approvals and rejections;
  - soil report and any related correspondence;
  - daily records including site diary, progress photographs, telephone logs, material receiving and OSD reports, labour, plant and equipment records, overtime records, stores requisitions etc;
  - shipping schedules, bills of lading, invoices and other documentation;
  - daily and monthly progress reports;
  - QA/QC documentation including test and inspection reports, weld records, etc;
  - RFI submissions and related correspondence;
  - change order requests, estimates, approvals and all related correspondence;
  - payment applications and certificates;
  - sub-contracts and all sub-contractor documentation, correspondence etc;
  - programmes and schedules and all revisions, updates etc;
  - hand-over/taking over and completion certificates;
  - punch list;
  - close out documentation.
- Thoroughly read through all documentation to establish both principal and ancillary claim issues.
- Identify items pertinent to the claim issues to be pursued.
- Extract and record from documentation the data relevant to each claim situation.

- Establish claim folders for each claim situation, containing pertinent data arranged chronologically.
- Establish responsibility and liability for each claim issue.
- Prepare draft narrative outline of each claim issue putting forward case for argument.
- Prepare preliminary estimate of time and cost effects relative to each claim issue.
- Discuss issues with project team, visit site and get feel for the project (if not already familiar with it).
- Establish merit of each claim issue.

Step 2 – Schedule analysis

- Review and compare as-planned and as-built schedules, in whatever form available.
- Summarise all schedule changes, revisions and updates with dates, reasons etc.
- Assess whether as-planned schedules were realistic when prepared.
- Create impacted schedule showing affects of changes, revisions and updates.
- Prepare fragnets[3] for each claim issue.
- Identify time delay periods, acceleration periods and disruption periods.
- Calculate delay days due to contractor and delay days due to others.
- Identify compensable, non-compensable, excusable and non-excusable delays.

Step 3 – Cost analysis

- Compare actual hours against hours included in tender in relation to total project, areas concerned with claim issues, specific trades etc.
- Identify overtime hours worked on an area by area and trade by trade basis, with particular reference to areas and trades affected by claim issues.
- Identify plant, equipment, overhead and direct costs and compare against tender estimates.
- Determine cause of delays and/or extra costs on area by area and trade by trade basis with particular reference to areas and trades affected by claim issues.
- Determine responsibility and liability for delays and/or extra costs.

Step 4 – Calculations

- Determine delay damages and acceleration costs.
- Determine costs of disruption (loss of productivity etc.).
- Determine costs of all changes.
- Determine prolongation costs – overheads, liquidated damages etc.
- Determine loss of interest on capital used due to delays, additional work, late payment etc. where not included for elsewhere (such as in new rates for change orders).

- Prepare claim summary listing all claims for additional costs, damages, liquidated damages etc.

Step 5 – Preparation of claim document

- Prepare claim document under established heads:
  - Cover sheet;
  - Index;
  - Executive Summary;
  - List of Claims;
  - Claim Analysis (for each claim):

  - Summary of Claim – brief description of basis for the claim and amount sought;
  - Underlying Facts – brief summary of relevant facts supporting claim;
  - Entitlement Analysis – rationale behind contention that contract allows for the compensation sought; analysis applying the relevant facts to the contract provisions and any other legal basis for prevailing on the claim;
  - Quantum Calculation Analysis – set forth the exact amount of additional costs being sought including the calculation used as the basis for the amount claimed (for example, Head Office Overhead claim could use either Eichleay's formula,[4] Emden's formula[5] or Hudson's formula[6]);
  - Factual Support of Claim – including:
    - contract provisions which support both the entitlement and the quantum calculations;
    - chronology of events and correspondence presented in a columnar format indicating:
      - when the claim work was discovered;
      - when the claim work was reported;
      - when the claim work was started;
      - when the claim work was completed;
  - list of documents relating to the claim including:
    - contract documents;
    - specification;
    - drawings;
    - relevant correspondence;
    - clarifications and Requests for Information schedules;
    - change order and RFI logs;
    - photographs;
    - schedules;
    - graphs and charts supporting and clarifying contentions and records;

  - any other relevant documents;
  - footnote summaries;
- Attachments – each claim should be supported by separate copies of the relevant contract and specification clauses, drawings, clarifications/requests for information, schedules, job diary references, RFI and change order logs, submittal logs, test records, correspondence and any other information relevant to that particular claim.

- Prepare exhibits to support claim presentation.
- Review and finalise claim document with project team.
- Discuss and review claim document with company management and agree strategies for presenting it to other party.
- Brief legal representatives if appointed.

A well prepared and presented claim which includes full supporting data, records, photographs and so on and a concise, well-written narrative and statement of claim based on irrefutable facts will be better received than a half-hearted effort with a poor standard of supporting data, badly presented and with little if any substance. It cannot be over emphasised that content and presentation are all important and if these are not given the maximum of attention then the probability of success must be seriously decreased.

1. See 'Claims and the Arbitrator' section on p.51.
2. CPN is an acronym for 'Critical Path Network'.
3. A 'fragnet' in programming terminology is a subdivision of a project network diagram usually representing some form of sub-project or, in a claim context, the sequences of work preceding, during and succeeding and relevant to a particular claim issue.
4. Eichleay's formula: Stage 1 – (Contract Turnover ÷ Company Turnover) x (Total Overheads for Contract Period ÷ Contract Period) = Daily Overhead Contribution for project. Stage 2 – Daily Overheads Contribution x Delay Days = Amount Claimed. (Favored in the US and other International environments.)
5. Emden's formula: Actual Head Office Overheads and Profit Percentage x (Contract Sum ÷ Contract Period) x Period of Delay = Amount Claimed.
6. Hudson's formula: Head Office Overheads and Profit Percentage included in Tender x (Contract Sum ÷ Contract Period) x Period of Delay = Amount Claimed *(N.B. – the inclusion of profit assumes that the profit percentage included in the tender was both reasonable and earnable)*.

Claims and the Arbitrator

It is important, when preparing a claim for presentation to the other party, to bear in mind, as mentioned in the previous section, the more than likely possibility that the dispute in question will end up being resolved through a form of ADR or arbitration. It is advisable to prepare the claim in such a way that its chances of success are enhanced to the fullest extent possible, basing the cause, effect and quantum of the claim as well as its general content on provable facts and contract clauses accompanied by as much supporting evidence in the form of contract documentation and contemporary records as it is possible to put together.

Arbitrators will be chosen for their particular expertise in one or more aspects of construction or in the law applicable to the particular contract. The arbitration tribunal might be a mix of disciplines or may be composed of members of the same discipline. The likely mix of disciplines making up the tribunal should be found out before preparation of the claim, if at all possible at that stage, in order to ensure that it can be tailored to suit. There is no need to go into detail on the very basics of an engineering issue if the tribunal is composed of engineers; conversely, if the tribunal is composed primarily of lawyers then they will need as much guidance as is necessary to enable them to understand the relevance and the complexities of the engineering issues in dispute. A panel with no legal expertise will need comprehensive guidance on legal issues relating to the claim whereas a panel of lawyers will not.

If the arbitration is being held in a country other than the claimant's home country it must ensure that any translation required is performed by translators of the highest repute and that they are conversant with both the technical and the legal jargon they are being asked to translate.

It is essential that the claim issues can be tied to contract clauses. There should be ample references in the claim presentation to the various contract clauses that are relevant not only to each specific issue in dispute but also to the contractor's case as presented in the claim. This will help the arbitrators in their understanding of the issues in dispute and will ease their job in deciding the relevant merits or demerits of the claim issues. Without merit there can be no damages but there must not be a mixing of arguments of merit with arguments on damages. Merit must be established first before pursuing the quantum of damages, and merit must come from the contract.

One of the intentions paramount in the minds of those opting for an arbitration hearing is to get a resolution of a dispute in the minimum amount of time. To this end it is superfluous to call too many experts to testify on a specific point. On all but the largest project or the most complex claim, one expert witness should suffice in respect of each specialist topic as long as that expert's opinion is presented in a clear and concise manner. Similarly evidence should be kept to the minimum and attention should be paid to only presenting that which is irrefutable or at least has a high credibility rating. The presentation of large amounts of questionable, unsupportable, frivolous and irrelevant evidence can only be counterproductive.

Where there are several claim issues it is advisable to separate the arguments affecting each claim and to fight each on its own particular merits. There are going to be times where claim issues overlap and if this is the case then a combined approach would be acceptable and probably unavoidable; however, care must be taken to establish the individual merits of each in as clear and concise a way as possible.

The quantification of the damages claimed must follow four basic guidelines:

- the type and quantum of damages claimed must be credible – it is not wise to include any 'tongue in cheek' claims as once discovered these will put a question mark over the whole claim and have a damaging impact on its outcome;
- damage claims must be directly related to the project works and, wherever possible, linked to contract clauses. Claims that are secondary to the contract may be more difficult to justify and may even be outside the arbitrator's authority to decide on;
- damage claims must be based upon provable facts. A claim issue should only include matters directly related to that issue and should not be used as a vehicle to resolve other unconnected issues;
- the supporting data made available must be sufficient both in terms of quality and quantity to prove the amounts of damages claimed. Actual records made at the time of the claim event, including minutes of meetings, invoices, photographs as well as correspondence, will strengthen the claimant's case whereas a lack of any such records will significantly weaken it.

A good award must be clear, concise and definite in its decision and in the actions it deems necessary for the parties to undertake in order to comply with its ruling. The award should decide every issue placed before the arbitrator and should not rule on anything outside the scope of the arbitration or the arbitrator's authority. These criteria must be considered at each step of the claim preparation process so that there are no obstacles put in the way of a good award. The claimant has a responsibility to ensure that the claim is presented clearly and concisely, is supported by adequate records, is directly related to the contract and is within the arbitrator's authority as established in the contract.

In concluding this section it is emphasised again that the contract/claim management process begins when the contract is signed and all efforts should be made to maintain clear and concise records of all events and occurrences in connection with the contract works in case a claim situation should develop at a later date. When such a situation does subsequently develop, particular care should be taken to maintain records directly related to the claim event. These would include records relating to the events leading up to the claim event as well as those raised during and subsequent to the event and relating directly to it and to its effects on other aspects of the contract. The success of a future arbitration will depend upon this. As stated many times before, there is no substitute for accurate contemporary records when substantiating a claim.

Concurrent Delay

Concurrent delay is when two or more delaying events occur and, even if one of the delaying events had not occurred, it is obvious that the schedule would have been delayed by one of the other events. In the strict definition[1] of concurrent delay the events should occur at precisely the same time but in practice this is rarely the case.

Where the events occur at different times but the effects of them are felt at the same time this is better referred to as the concurrent effect of sequential events (such as a change order issued in week 10 and late information in week 15 resulting in delay at the same time in week 24). It can also be said that two unrelated delay events are deemed concurrent if they fall on parallel critical paths but would not be deemed concurrent if either one was, or both were not, on a critical path, although in such a case they might ultimately be deemed concurrent if they exceeded the total float available in those paths.

Where the contractor has already been granted an extension of time and then incurs a problem of its own making which, but for the granted extension of time, would have delayed completion but which does not add to the existing delay, it is protected from the consequences of the second delay. The extension has already been granted and cannot be taken away.

If the contractor is delayed by two events at the same time, one of its making and one not of its making but which would have entitled it to an extension of time under the contract (a 'relevant event'), the contractor would be entitled to an extension of time for the period of delay caused by the relevant event.[2]

Where the contractor is in delay because of an event which does not entitle it to an extension of time and then a relevant event occurs which would have delayed it had it not already been in delay, and that event would have entitled it to an extension of time, the contractor will argue that it has an automatic right to an extension of time. This is not looked upon kindly by the courts, who have ruled that where a relevant event occurs during a period when the contractor is already in delay due to its own making, if the works are not likely to be delayed further as a result of the relevant event occurring then it has no effect upon the completion date and need not be considered.[3]

The relationship between the critical path and the events is an important consideration and must be addressed in all delay cases. The critical path should be redrawn at the point in time immediately preceding the relevant event in order to judge with a degree of accuracy its effect on the scheduled work.

The Canadian courts apply the practice of 'apportionment' when allocating damages for breach of contract[4] and this approach is also gaining acceptance in New Zealand.[5] In apportionment, the courts, or tribunals, spend time allocating the causation of the delay and the monetary consequences between the employer and the contractor.

This approach has not found favour with those responsible for resolving construction disputes in England and Wales, or in fact in most other common law jurisdictions. In these jurisdictions the 'all or nothing' approach to the application of the principles of causation is preferred (i.e. the courts prefer to attribute delaying events to a single cause resulting in the plaintiff succeeding either completely or not at all). In the instances where apportionment has been put forward to the English courts for consideration it has generally been rejected as an approach incompatible with the judicial principles of England and Wales.

Courts in the US have a different understanding of the term 'apportionment' when applied to two or more causes of delays in that to them it does not refer to the sharing of responsibility for the period of delay but to the tribunal's actions in investigating and allocating different parts of a delay period to different causes.

The US separates delays into:

- excusable delays;
- excusable/non-compensable delays;
- excusable/compensable delays; and
- inexcusable delays.

Generally when delays are concurrent or intertwined, US courts will deny both parties any financial recompense unless either party can prove conclusively the delay upon which its claim is based.[6] Where there is an excusable delay concurrent with a non-excusable delay, US courts will grant an extension of time but will not award financial recompense. Similarly, when there are compensable delays concurrent with non-compensable delays, US courts will grant an extension of time with no financial recompense unless the contractor can show that that the compensable delay was the sole cause of the financial losses that it incurred.

The contractor need not address every possible delay issue in its claim presentation as long as it has sufficient evidence to support its position in its claim for employer-caused delay and as long as that evidence focuses on critical path delays. Once the contractor has submitted its claim, the burden then shifts to the employer who then has to show that, even without the introduction of the employer-caused delays, the project would have been delayed by the contractor-caused delays.

The employer must prove convincingly that the events of contractor-caused delays not addressed by the contractor in its claim submission nevertheless have relevance and that omission of these events affects the causative assessments. The employer must present a strong case sufficient to convince the courts that the contractor-caused delays should be the ones upon which the dispute should be determined and that the employer-caused delays were subsidiary in this respect (for example, to argue that they commenced after the contractor-caused delays were already in being).

The final decision as to how to approach concurrent delays in different jurisdictions will depend to a large extent upon the precise wording of the extension of time and loss recovery clauses in the particular contract concerned and upon the laws in the particular country concerned.

1. The UK's Society of Construction Law's 'Delay and Disruption Protocol' published in October 2002 (amended in March 2003) states in Guidance Note Guidance Section 1.4.2: 'Concurrency is a contentious issue both because there are differing views on the correct approach to concurrency when analysing entitlement to EOT [*extensions of time*] and because there are differences about the meaning of concurrency itself'.
2. *Henry Boot Construction (UK) Ltd* v *Malmaison Hotel (Manchester) Ltd* [1999] 70 Con LR 32, QBD (TCC).
3. *The Royal Brompton Hospital NHS Trust* v *Frederick Alexander Hammond & Others* (2002) (TCC).
4. *Tompkins Hardware Ltd* v *North Western Flying Services Ltd* (1982) 139 DLR (3d) 329; *Ribic v Weinstein* [1982] 140 DLR (3d) 258; *Doiron v Caisse Populaire D'Inkerman Ltee* (1985) 17 DLR (4th) 660.
5. *Day* v *Mead* [1987] 2 NZLR 443, CA; *Mouat v Clark Boyce* [1992] 2 NZLR 559, CA.
6. *Blinderman* 695 F. (2d) 552 (1982), US Court of Appeals for the Federal Circuit, at p.559.

Consideration

The definition of and requirement for 'consideration' is probably well known to most readers, at least in general terms. However, it is worth refreshing ourselves on the basic principles in order that we can see how these might or might not have the same relevance in different jurisdictions.

For a 'simple'[1] contract to be legally binding under the laws of England and Wales, the US and most countries whose legal system is based on the English model of a Common law[2] system, it must be accompanied by consideration.[3] Consideration will be something of legal value done, paid, given or promised by the party seeking to enforce the contract. Consideration does not stand alone as the defining characteristic of a simple contract but will be accompanied by the other essentials of a valid simple contract, namely an agreement,[4] an intention to create legal relations,[5] capacity[6] and formalities.[7]

There are circumstances where consideration might be given as part of an arrangement without the parties having any intention of creating a legal relationship, usually in a social or domestic context, and as such each case must be looked at taking due account of the nature of the agreement and the circumstances by which it came into being. This will rarely if ever be the case in international construction contracts.

Consideration has been described as consisting either in:

> *some right, interest, profit or benefit accruing to one party, or some forbearance, detriment, loss or responsibility given, suffered or undertaken by the other.*[8]

Consideration may be a promise given in return for a promise, a promise given in return for an act of forbearance or an act of forbearance in return for a promise. When consideration is given in the form of a promise this is termed 'executory consideration' whereas consideration in the form of an act or forbearance actually carried out (such as finding a lost article for which a reward has been offered or positively refraining to enforce a debt) is called 'executed consideration'. The consideration cannot be 'past consideration'[9] but must be present when the contract is made.

The consideration need not necessarily represent an adequate return for what has been promised or done but must be good (legally sufficient) and have some degree of value in the eyes of the law. This perceived value might be in the form of permission to perform an act, an agreement to postpone an action or the surrender of a document that subsequently turns out to be worthless in itself. An exception to the value rule is where there is a contract between close relatives, where the consideration can be good even if not valuable because the love and affection between the relatives may be legally sufficient even if it cannot actually be valued in money terms.

A promise would not be considered to fulfil the requirements of valuable consideration if:

- the act or promise was one which the promisor was in any case bound to give by law; or

- the act or promise was one which the promisor was already bound to do and was made to the person to whom he was already legally bound; or
- the act or promise represented an illegal act or was contrary to public policy.[10]

Payment of a lesser sum would not be acceptable as consideration for the satisfaction of a promised larger sum,[11] although if the parties actually agreed to accept a different consideration from that originally promised, even if that consideration was of a lesser value than the original, this would be acceptable and the contract would be said to be discharged by 'accord and satisfaction'.

As an example, a manufacturer sold a piece of equipment to a client who intended installing it in his factory. Due to a recession in the industry the client cancelled his new factory and asked if he could sell the piece of equipment on before the supplier had installed it. This was not a situation covered by the supply and install contract.

The supplier agreed on condition that complete confidentiality was maintained regarding the advanced technology used in the manufacture of the equipment and that the supplier had power of veto over the company to whom the equipment details could be released. A separate confidentiality agreement was drawn up to cover these requirements including the two original parties plus the prospective purchaser as signatories.

This was not a contract by deed (as was the existing contract) nor was it a variation to the existing contract. It was in fact a simple contract between the two parties to the original contract, including also the prospective purchasers. The consideration in this case was the supplier's promise that its technology details could be shown to prospective purchasers in return for the other parties' promises to keep such information confidential. This was an actual situation encountered on an international construction contract involving parties of three different nationalities working in three languages in three different countries.

Consideration is not, however, necessary for a contract to be binding in Civil law countries. In France, for instance, the four requisites under the French Civil Code that are essential for the validity of agreements[12] are:

- 'The consent of the party who binds himself.'
- 'His capacity to contract.'
- 'A definite object which forms the subject matter of the undertaking.'
- 'A lawful cause in the obligation.'

When comparing the French Civil Code system to the English Common law system it is often said that 'cause' under the Civil Code takes the place of the 'consideration' necessary under Common law jurisdictions, and this is largely the case. The Civil Code states in relation to cause[13] that:

- 'An obligation without a cause or with a false cause, or with an unlawful cause, may not have any effect.'
- 'An agreement is nevertheless valid, although its cause is not expressed.'
- 'A cause is unlawful where it is prohibited by legislation, where it is contrary to public morals or to public order.'

In Japan under the Civil Code of the country no formality is needed to create a valid construction contract, merely a meeting of minds. The Japanese Construction Business Act (1949, amended 1971)[14] provides directory provisions.[15] Although they cannot pre-empt the Civil Code, the directory provisions mandate that parties to a construction contract must write down the following matters in a contract and must either append their signatures or their names and seals:

- description of works;
- price of works;
- construction start and completion times;
- when and how advance and progress payments are to be made;
- how to compensate for changes in construction schedule or contract amount and for design changes, contract suspension or termination;
- sharing and evaluation of loss in respect of force majeure events including Acts of God;
- changes in contract amount or scope due to changes in materials or services;
- sharing of third party damage liability;
- method of delivery of owner-supplied materials or owner-rented tools or equipment;
- times and methods of inspection of wholly or partially completed construction and time of delivery of completed construction;
- time and method of payment upon completion of construction;
- interest, penalty or other damages in case of delay in performance of contractual obligations and other liabilities;
- method of dispute resolution.

While discussing consideration and, incidentally, the formation of contracts, it might be worthwhile touching on an expression which is in common usage internationally but which those not previously involved in international construction might find strange. The expression is 'Counterpart Execution'.

A counterpart is a copy or a duplicate of a document. In contractual terms, a counterpart can be a copy or duplicate of an agreement, contract, licence etc. Unlike domestic contracts in England and Wales, the management of international contracts frequently involves players from many different countries having to enter into contracts or sub-contracts with each other often while physically located thousands of miles apart. In such situations it is often inconvenient to execute a document that requires the individual signatures of each party on the same copy of the document and it is more expeditious for each party to sign only the particular copy in its possession.

When each party has signed their respective copies (counterparts) of the document, be it an agreement, contract or licence etc., it is deemed to have been executed and the separate counterparts shall be considered as one complete executed document. They are binding so long as all the requirements of consideration or validity of agreement have been fulfilled.

1. 'Simple' contracts include all contracts not made by deed (under seal – also known as a speciality contract). Simple contracts may be oral (parol contracts), written (contracts under hand) or formed by conduct (inferred contracts).
2. This is a requirement under the laws of England and Wales but not under Scottish law nor in many other countries which have a Civil Code legal system – See Appendix 2: International Legal Systems on p.269.
3. Consideration must have some sort of value and is usually referred to as 'valuable consideration' – in this section we shall refer to it as 'consideration' in the interests of brevity.
4. Agreement – consensus ad idem (consent to the same thing).
5. Shown either by the form in which the agreement is expressed or by the giving or promising of consideration by one party in return for the benefit claimed from the other party.
6. Parties must be legally capable of incurring the contractual obligation which the agreement purports to impose upon them.
7. Examples of contracts which are required to be in writing but not necessarily under seal (by deed) are Bills of Exchange, promissory notes, contracts of marine insurance, certain categories of finance/loan agreements, transfer of copyrights and as required by statute.
8. *Currie* v *Misa* (1875) LR 10 Ex. 153.
9. Executed before the promise was made.
10. In this case only the objectionable provision would be void, the remainder of the contract would stand unless the voided portion represented the whole or the major part of the consideration.
11. There might, however, be good discharge in Equity in accordance with the 'High Trees House' principle from *Central London Property Trust Ltd* v *High Trees House Ltd* (1947) KB 130 which states that where a promise is made which is intended to create legal relations, is intended to be acted on by the promisee and is in fact so acted upon, the promise will bind the promisor to the extent that he will not be allowed to enforce a prior contract which is inconsistent with it, although he could still not be sued upon it since there is no consideration and hence the promisee would have no right of action.
12. French Civil Code, Article 1108.
13. *ibid.*, Articles 1131, 1132 and 1133.
14. Kensetsu-gyo Ho.
15. Article 19 – Contents of Construction Contracts.

Contemporary Records

It is common for conditions of contract to require contractors to keep contemporary records that can be used to support any claim contention on their part.[1] It is in any case good policy for contractors to maintain good records at all times as these will prove themselves to be invaluable not only as a means of supporting their own claims but also of defending claims made by the employer or sub-contractors.

A competent contractor would be considered foolhardy should it decide against keeping adequate records merely because it had no intention of claiming against an employer. Even if that was to be its policy, the contractor could have no idea at the stage when the policy was introduced as to what claims might be made against it at a future date. This is, however, an attitude frequently adopted by contractors from the Far East, in particular contractors from Japan where their inherent reticence in dealing with foreigners makes them wary of initiating claims (or even quoting contract clauses in the simplest of correspondence) for fear of 'offending' an employer, damaging their then current relationships and damaging their chances of further work. They have no need for keeping such records in their domestic business environment and so this 'head in the sand' approach to international contracting saves them the effort of retraining their staff into the ways of the larger international contracting community.

Unfortunately by adopting this attitude they lose out not only in being unable to defend claims against them but also in being unable to put together comprehensive claims even when they have a cast iron case in principle. The not insignificant number of Western engineering, contract and claims specialists employed by Japanese companies are making inroads into this attitude but still have a long way to go, due in no small part to the fact that although the senior international managers of the majority of Japanese companies trading overseas are aware of the need for contractual direction the 'salary men' in Japan who are largely responsible for maintaining the records are reluctant to act on the advice of a 'gaijin'.[2]

Questions have been raised as to what represents 'contemporary' records and what happens when such records are not submitted in due time but could be substantiated by verbal confirmation from personnel directly related to the contract works.

'Contemporary' in this context has the meaning of 'coming into being during the same period of time as' the event or occurrence to which the records relate. This would mean in a situation where the contractor was claiming an extension of time due to a particular event or circumstance, records should be kept from the start of the event or occurrence until the completion of the works which resulted from it or of the delay which it caused.

It has been held that not more than a few weeks should elapse between the works being carried out and the production of suitable records. It was also held in a case concerning the Falklands Government that, where the records for one week were missing from a series of records covering several weeks, verbal evidence confirming records for the missing week would not in itself be acceptable.[3] In that particular case the judge held that the contractor could rely on a witness statement **only** if the witness statement either:

- illustrated how and when the contemporary record came into being; or
- sought to clarify any ambiguity in the contemporary record.

The contractor could not rely on the witness to 'plug the gaps' in the records which it should have been keeping at that time.

A contractor who fails to meet its contractual obligations to provide sufficient contemporary records in support of its claim will not be allowed to adduce evidence at a later date to supplement those insufficient contemporary records where such evidence would put the contractor in a better position than it would have been in had it in fact complied with its contractual obligations. It will, however, be able to clarify any ambiguity created by a contemporary record that is unclear or otherwise inadequate because it had been recorded in a hurry without being afforded the level of care that it should have been afforded.

A 'record' in a construction contract context has been defined as being 'original or primary source' material.[4] A document, irrespective of its form, will only be a record if:

- it effects a transaction or is a contemporaneous compilation of information derived from primary sources which is intended to serve as a record of events; and
- it is a comprehensive compilation, rather than a selection of source information.

If a document is not a comprehensive record of what has occurred, or, even if comprehensive, was not intended to serve as a primary source of information on that matter, it will not be a record.

Original or primary source records would be, for instance, time sheets completed by site workers at the very time that they were carrying out the work concerned. Where they failed or were not required to fill out such timesheets, the contractor's site representative's records completed at the time recording the number of workmen carrying out that work would be classed as original source records. A notation prepared some weeks or months after the event by a clerk at the contractor's head office from second or third hand information would not be classed as contemporary records in this instance.

It cannot be stressed strongly enough that where contemporary records are required to be kept these must be kept at the time that, or soon as possible after, the event giving rise to the claim occurs and subsequently during the entire length of its currency in as clear and unambiguous a fashion as possible. As with all other matters relating to the proper control and management of claims, it is essential to conform to the requirements of the contract in every way in order to preserve the right to claim and the means of justifying the facts and the quantum of that claim. Where there are no contemporary records the claim will in all probability fail, or will at least result in a diminution of the claim in time or cost terms, depending upon the exact wording of the relevant clause in the contract upon which the claim relies.

The position under English law is clear in that the requirement to keep contemporary records is a strict one and the admittance of supplementary evidence to overcome a want of contemporary records would not be allowed except in the very narrowest of circumstances. It is difficult to know with any certainty, however, whether Civil Code or Middle Eastern jurisdictions would follow such a strict interpretation since they tend to put greater weight upon the importance of personal witness testimony than the contemporary records favoured by common law jurisdictions. Nevertheless there cannot be any harm in keeping a comprehensive collection of contemporary records as even if witness testimony attracts greater interest generally under a particular jurisdiction, the party that has a substantial and convincing collection of contemporary records as evidence to support its case will surely gain considerable status in the eyes of the arbitrator.

1. For example, FIDIC 'Red Book', 'Yellow Book' and {'Silver Book'} 1999 Edition Clause 20.1 '...The Contractor shall keep such contemporary records as may be necessary to substantiate any claim, either on the Site or at another location acceptable to the Engineer {Employer}...If the Contractor fails to comply with this or another Sub-Clause in relation to any claim, any extension of time and/or additional payment shall take account of the extent (if any) to which the failure has prevented or prejudiced proper investigation of the claim....' (similar clause in ICE Conditions of Contract 7th. Edition).
2. Japanese name for a foreigner.
3. *Attorney General for the Falkland Islands* v *Gordon Forbes Construction (Falklands) Ltd (No. 2)* (2003) (LTL 14/3/2003) in relation to Clause 53.4 of FIDIC Conditions 4th Edition, 1987.
4. Defined in *H* v *Schering* (1983) 1 All ER 849.

Contra Proferentem Rule

The doctrine of 'contra proferentem' is used where there are perceived ambiguities or discrepancies within or between documents. The words are actually a shortened form of the Latin phrase 'verba chartarum fortius accipiuntur contra proferentem' which means literally 'the words of written documents are constructed more forcibly against the party offering them'.

In practice the application of the doctrine of contra proferentem serves to place the risk of ambiguity upon the drafter, i.e. where there is ambiguity in a document or between various contract documents, the words are construed against the party putting forward the document(s), being the party that prepared them or at least ordered and concurred with their preparation, content and particular application.

Standard conditions of contract, such as FIDIC,[1] which do not generally set out to favour one party more than the other and which are not imposed arbitrarily by one party upon the other, are not generally open for application of the contra proferentem rule; however, in certain instances, such as the application of liquidated damages and extension of time clauses, it has been held by the courts to apply.[2] It is not unusual for courts to be wary of potentially controversial clauses such as liquidated damages, exemption and forfeiture clauses even in standard forms of contract and in such cases they are often willing, indeed keen, to apply the contra proferentem rule. This might seem a little strange given the degree of involvement of representatives of both employers' and contractors' organisations in the drafting stages of such documents but it is still, nevertheless, a fact of life given added weight by the fact that it is still usually the employer who states at tender stage which forms of contract will apply.

From a contractor's viewpoint the contra proferentem rule is therefore an approach to be considered when questioning the application of contract clauses/terms/conditions, the liquidated damages clause in particular. If there are sufficient grounds for proposing this as a means of restricting the application of the liquidated damages clause then the courts, arbitration tribunal or whatever would be inclined to look favourably upon such an argument.

In a case in a Local Division of the High Court in South Africa[3] the judges noted that an insurance company's general conditions referred to a claim being fraudulent but did not refer to a claim *form* being fraudulent – they held that since the insurance company was the author of the policy, where such ambiguity arose the contra proferentem rule should operate. In a Canadian case, *Squire* v *Insurance Corp. of British Columbia (BCCA)*,[4] the court expressed the view that the contra proferentem rule has no application to the statutory terms of a contract of insurance. The same principle would seem likely to apply to any statutory term included in or incorporated into any form of contract or agreement, including those used in the international construction industry.

In a contract which is not based upon standard conditions there is even more scope for the application of this rule. The contra proferentem doctrine has, in addition to insurance claims, been used in relation to claims involving the interpretation of contractual, commercial and technical documentation, procurement and shipping arrangements to name but a few.

The German Civil Code in its revised 'Law of Obligations' makes particular mention of 'surprising and ambiguous clauses'[5] and actually codifies what is in effect the contra proferentum doctrine in respect of the use of standard business terms stating:[6]

> *In case of doubt, standard business terms are interpreted against the user.*

The French Civil Code states:[7]

> *In case of doubt, an agreement shall be interpreted against the one who has stipulated, and in favour of the one who has contracted the obligation.*

This again confirms the application of the 'contra proferentem' maxim in both Common law and Civil Code systems.

It is unusual to find a clause in a construction contract excluding the application of the contra proferentem rule but such clauses have been found in certain ancillary documents, such as licensing agreements, technology transfer agreements and the like, and in agreements which have been through a process of negotiation by and between the contracting parties. There is no reason why such a clause could not be included in a construction contract or agreement under similar circumstance, i.e. where the contract or agreement was entered into following a period of detailed negotiation by the contracting parties, but to include it under any other circumstances would seem unfair and against normal industry custom.

The German Civil Code actually prevents the exclusion of the application of the 'contra proferentem' rule (noted above as confirmed by its Article 305(c)(2)) and states:[8]

> *The rules in this section apply even if they are circumvented by other arrangements.*

1. Fédération Internationale des Ingénieurs-Conseils (International Federation of Consulting Engineers), issuers of standard forms of contract for use in various international construction situations, including the Red Book, Silver Book and Yellow Book.
2. *Peak Construction (Liverpool) v McKinney Foundations Ltd* (1970) 1 BLR in which Lord Justice Salmon stated that 'The liquidated damages and extension of time clauses in printed forms of contract must be construed strictly contra proferentem.'
3. *Strydom v Certain Underwriting Members* (2000) (2) SA 482 (W).
4. *Squire v Insurance Corp. of British Columbia (BCCA)* (1990) 69 DLR.
5. German Civil Code, Article 305c.
6. *ibid.*, Article 305c (2).
7. French Civil Code, Article 1162.
8. German Civil Code, Article 306a.

Contract Close-Out

'Contract close-out' is an expression not commonly used in England and Wales but those involved in international construction, particularly of an engineering nature, will be familiar with it. Basically it is similar to what would be called the 'Final Account' stage in England and Wales but more comprehensive in nature. It not only involves the finalisation of any necessary re-measurement, adjustment of change orders, provisional sums an so on but also the coordination and finalisation of all documentation, including that noted below, without which the final certificate cannot be issued. This applies between the contractor and its sub-contractors as well as between the contractor and the employer. This section has been written in the context of the contractor closing out its sub-contracts but could equally apply in principle to the employer closing out with the contractor.

Contract close-out between the contractor and its sub-contractors should start in earnest at the time of Taking Over (often also referred to as 'Provisional Acceptance' or 'Substantial Completion'), when the sub-contract works have been completed and the Defects Liability (Warranty) Period commences, as evinced by the issuance of a Taking Over/Provisional Acceptance/Substantial Completion Certificate,[1] and ends upon the issuance of the Final Acceptance Certificate.[2] This is the period when the contractor has to ensure that all the documentation which it expects and requires its sub-contractors to deliver to it prior to the issuance of their Final Acceptance Certificate has been submitted in the form, style and quantity which the sub-contract calls for it to be in.

The most common documentation required at this point in time could consist, without limitation, of:

- as-built drawings;
- QA/QC records (including test results, actual performance values etc.);
- operation and maintenance manuals;
- warranties and guarantees from sub-vendors; and
- punch lists.

The sub-contractors will also have to satisfy the contractor that they have fulfilled all their sub-contract obligations, have complied with all necessary legal requirements of the sub-contract, have provided all spare parts required of them, have completed any training programme required of them and have settled all their sub-vendor and other payments which they were obligated to make under and in connection with their obligations under the sub-contract.

The latter must be evinced by receipts from the sub-vendors etc. releasing the contractor from any further liabilities in this respect. In some countries, particularly in the Middle East, it is common for the contractor to have to provide evidence to confirm that it has paid its employees their wages and salaries in full and also that it has made all necessary contributions to the local social insurance offices on behalf of its employees. On one particular project I worked on in East Africa we even had to provide receipted hotel bills for the expatriate staff members employed on the project before our client (the ruling government at the time) would finalise our account.

If there are any outstanding measurement, pricing or claim items at the time of close out, provisions must be made to ensure that procedures are in place to resolve these in a manner that is to the satisfaction of all the parties. Measurements and pricing issues should be relatively easy to resolve where the sub-contract makes provision for either re-measurement or for some other means of evaluation of changes to the works (such as dayworks, agreed lump-sum quotations etc.). Where there are unresolved delay or other claims this will inevitably take a little longer to resolve and any which are disputed could well end up being referred to arbitration, although hopefully the contract will provide for a particular method of ADR to be undertaken prior to invoking the arbitration clause.

The parties must ensure that whatever the sub-contract calls for as regards close-out in these circumstances is put into action within the timescales and in the manner set down in the sub-contract and they must take great care to ensure that no claims are time-barred because of their lack of diligence. All future actions that survive[3] the completion of the work must be identified and recorded appropriately.

When the sub-contractor's Final Completion Certificate is issued it must state clearly whether it confirms either that all outstanding issues between the parties have been resolved or whether there are still any to be resolved after its issuance. There might, for instance, be an on-going arbitration hearing that will continue for some time after the Final Completion Certificate is issued and this must be noted as such in the Final Completion Certificate.

The contractor should prepare a Contract Close-out Check List[4] and should obtain the sub-contractor's agreement and signature when the items listed on it are closed to their mutual satisfaction.

As mentioned earlier, similar procedures would apply in principle between the employer and the contractor although the items listed on the Close-out Check List would in all probability have to be modified to suit the particular contractual requirements of the main contract. Each party must, in that case, ensure that each of its various departments, consultants etc. is fully aware of what is required of it in this respect, and the deadlines which they have to work to, in order to avoid any delay to the employer's issuance of the Final Certificate to the contractor.

1. See sample Sub-Contractor Taking Over Certificate in Appendix 9 on p.301.
2. See sample Sub-Contractor Final Completion Certificate in Appendix 11 on p.305.
3. Typical obligations that would survive completion or termination would include such clauses as:
    (a) applicable and governing laws;
    (b) confidentiality agreements;

(c) dispute resolution procedures;

(d) guarantees and warranties;

(e) indemnities;

(f) language;

(g) limitation of liability;

(h) patent and intellectual property rights;

(i) notices.

4. Typical 'Sub-Contract Close-Out Check List' between contractor and sub-contractor is shown in Appendix 10 on p.303 – a similar format would be used between contractor and employer, adapted as necessary.

Contract – Maxims and Rules of Interpretation

There will be situations during the administration of the contract when questions will be asked regarding the true meanings and intentions of certain words or phrases contained within the contract. All parties will need to be conversant with the basic maxims and rules in use internationally for interpreting the construction of statutes and for determining the true meaning of particular or contradicting clauses in a contract.

Intention is to be found from the contract itself

The intentions of the parties must be able to be ascertained from the contract itself. The courts in Common law countries will not, in construing written contracts, go outside the written documents in order to substitute the presumed intentions of the parties where the presumed intentions cannot themselves be ascertained from the contract itself. In Civil Code jurisdictions the approach is somewhat different. Whereas English courts will try to find a solution from the contract terms, whether they be written or implied, the French courts will look to the nature of the contract itself.

The French Civil Code states that:[1]

> *one must in agreements seek what the common intention of the contracting parties was, rather than pay attention to the literal meaning of the terms.*

The French Civil Code goes one step further in interpreting contract clauses and states[2] that:

> *Terms which are customary shall be supplemented in the contract, even though they are not expressed there.*

Contract to be read as a whole

The intention of the parties to a contract is to be deduced from a study of the contract as a whole and, as far as is possible, by giving effect to each of the provisions contained within it. The court will first have to decide which documents actually form the contract between the parties and then decide what order of precedence is to be applied to the application of these various documents. The order of precedence may or may not have been expressed in the contract or by the parties in negotiations leading up to its formation.[3] This approach is similar in both Common law and Civil Code systems and is often confirmed by a clause to this effect written into the contract.

Ordinary meaning

Words written in a contract are to be given their ordinary, plain meanings as generally understood, unless to do so would give rise to inconsistency or absurdity. In trade or technical contracts or sections of contracts, the customary meaning of the words concerned as used in the everyday dealings of such trade or technical context shall be applied. English courts will find the natural and ordinary meaning of words but not so as to 'flout business commonsense'.[4]

Validity of meaning

Words, phrases or clauses capable of being interpreted in different ways or of having different meanings will be construed in Common law jurisdictions in such a way as to make the contract valid rather than to make it void or ineffective.

In a similar vein, the French Civil Code states[5] that:

> *where a clause admits of two meanings, one shall rather understand it in the one with which it may have some effect, than in the meaning with which it could not produce any.*

Particulars prevail over standards

Written words on a printed form will have greater effect than printed words, unless the contract documents expressly provide otherwise. The provisions in a written document will prevail over provisions in any incorporated documents. Following on from this, special conditions, or particular conditions written for a particular contract, will generally prevail over standard conditions unless the order of precedence specifically states otherwise. Again, there is often a clause written into the contract that confirms this intention.

Words will prevail over figures

Measurements stated in written or printed words or letters will prevail over measurements noted in numerals on or to be scaled from drawings or the like. Again, there is often a clause written into the contract that confirms this intention.

'Expressio unius'

The full Latin expression from which this phrase has been extracted is 'expressio unius est exclusio alterius' which means that the express mention of one thing rules out other things not expressly mentioned. This has been interpreted in a construction contract context as meaning that 'the express mention of a certain thing will exclude other things of a similar nature where things of a similar nature are not expressly mentioned'. Thus, a contract for the sale of an industrial building plus two houses, together with fixtures in the houses, was held to exclude fixtures in the industrial building. It is a maxim widely applied in Common law jurisdictions and has been referred[6] to as 'a well accepted canon of statutory construction' adding that it has to be applied 'with great caution'.

The French Civil Code takes a different view of things in certain contexts and states that:

> *Where in a contract one case was expressed for explaining the obligation, it shall not be deemed that it was thereby intended to reduce the scope of the agreement which extends as of right to cases not expressed.*

The French and English approaches are not totally opposite and in the example quoted the French would in all probability take the same view as the English. The French approach is probably more manageable than the English approach that has been recommended to be applied 'with great caution', as noted above.

In construction contracts it is common to find phrases such as 'without limitation to' or 'similar to' qualifying clauses where the application of expressio unius might otherwise diminish the application of that clause. For example, in force majeure clauses where various events are listed which would qualify as events of force majeure, the inclusion of 'without limitation' in the preface to the event listing will preserve the right to classify other events as force majeure events where it would be reasonable to do so (for example, adding hurricanes to an event classification which already includes storm and tempest but not adding subsidence, which would be regarded as unreasonably remote from the event listing already included).

'Ejusdem generis'

This is another Latin phrase which means 'of the same kind' or 'of the same nature' indicating that where particular words are followed by general words, the general words are to be taken to apply only to things of the same kind or nature as the particular words. Hence, in a charter contract it was held that the words 'any other cause' in the phrase 'war, disturbance or any other cause' did not include ice but were restricted to events of the same kind as war or disturbance (such as terrorist attack).

The French Civil Code states[7] that:

> *However general the terms in which an agreement is phrased may be, it shall include only the things upon which the parties appear to have intended to contract.*

This is basically the same as the Common law concept.

'Contra proferentem'[8]

This is a shortened form of the Latin phrase 'verba chartarum fortius accipiuntur contra proferentem' which means literally 'the words of written documents are constructed forcibly against the party offering them', i.e. that the construction least favourable to the person putting forward an instrument should be adopted against him. Where there is an ambiguity in a document, the words are to be construed against the party who put forward the document.

This rule is expressed very clearly in the French Civil Code, which states:[9]

> *In case of doubt, an agreement shall be interpreted against the one who has stipulated, and in favour of the one who has contracted the obligation.*

This confirms the application of this maxim in both Common law and Civil Code systems.

The Literal Rule, the Golden Rule, the Mischief Rule and the Purposive Approach

In addition, in English law there are three rules that are used by judges to interpret statutes. These are the Literal Rule, the Golden Rule and the Mischief Rule. There is also the Purposive Approach that is used to interpret statutes involving European Community Law, being drafted very differently from English legislation and favouring a Civil Code approach rather than a Common law approach.

The Literal Rule simply states that the words in the statute must be interpreted in accordance with their plain and ordinary meaning. Sometimes this results in absurd conclusions as in the case where a person had been charged with impersonating a voter at an election.[10] The charge was based upon him impersonating 'any person entitled to vote' but since the person he impersonated was dead (and therefore unable to vote) he could not be convicted. The Literal Rule interpretation had failed to find the true intent of Parliament when they passed that particular piece of legislation.

The Golden Rule is similar and states that words should be used in their literal meaning only to the extent that to do so would not produce an absurd or intolerable conclusion. It can be applied in two ways:

- if there is an ambiguity in the words of the legislation, preference will be given to the meaning of the word which does not result in absurdity; and
- it is used to avoid an outcome which is absurd or obnoxious to principles of public policy (even if the word(s) have only one meaning).

The Mischief Rule[11] is used by judges to decide what mischief or harm the statute was intended to remedy and they will consider four principles in their determination:

- What was the common law before the legislation was passed?
- What solution was offered by Parliament?
- What is the true reason for the solution? and
- What was the mischief that the law did not solve?

In one case where it was applied a person was found under the influence of alcohol while riding a bicycle. The Licensing Act 1872 allowed for a person in charge of a 'carriage' whilst under the influence of alcohol to be arrested without a warrant. The court interpreted 'carriage' as including a bicycle since they considered the intention of Parliament when passing the law was to protect the public at large from drunken people in charge of modes of transport of all kinds.

The Purposive Approach looks outside the working of the statute and is used in the interpretation of statutes relating to European Community Law, which are primarily based upon Civil law rather than Common law principles. Where a company dismissed its workers one hour before the company was transferred to different owners, the employees claimed unfair dismissal.[12] Under regulation 5 of the Transfer of Undertakings Regulations, which is a statutory instrument implemented in England and Wales to give effect to an EC directive, transfer does not terminate employment 'immediately before a transfer' took place.[13]

The courts used the Purposive Approach to import into the statutory instrument the words 'or would have been so employed if he has not been unfairly dismissed before the transfer' in order to ensure proper implementation of the EC directive which was obviously intended to protect employees' rights.

It can be seen from the above that there are two contrasting views as to how judges should interpret a statute. Whilst the literal and golden rules look to the actual wording of the statutes, the mischief rule and the purposive approach have a wider remit.

Rules of construction v rules of law

It is worth noting the distinction between rules of construction and rules of law. Rules of construction are applied to enable the courts to ascertain the intentions of the parties, as expressed, and to give effect to those intentions. Rules of law are applied to bring the parties within the framework of established legal rulings and these rules apply even though the parties may have expressed and intended something contrary and application of the rules of law defeats that intention.

In the US, maxims are listed in the codified statutes of most states, and include:

(a) 'When the reason of a rule ceases, so should the rule itself.'
(b) 'He who consents to an act is not wronged by it.'
(c) 'No one can take advantage of his own wrong.'
(d) 'No one should suffer by the act of another.'
(e) 'He who takes the benefit must bear the burden.'
(f) 'For every wrong there is a remedy.'
(g) 'Between rights otherwise equal, the earliest is preferred.'
(h) 'No man is responsible for that which no man can control.'
(i) 'The law helps the vigilant, before those who sleep on their rights.'
(j) 'The law respects form less than substance.'
(k) 'The law never requires impossibilities.'
(l) 'The law neither does nor requires idle acts.'
(m) 'The law disregards trifles.'
(n) 'Particular expressions qualify those which are general.'
(o) 'That is certain which can be made certain.'
(p) 'Time does not confirm a void act.'
(q) 'An interpretation which gives effect is preferred to one which makes void.'
(r) 'Interpretation must be reasonable.'
(s) 'Things happen according to the ordinary course of nature and the ordinary habits of life.'

1. French Civil Code, Article 1156.
2. *ibid.*, Article 1160.
3. See 'Ambiguities, Conflicts and Inconsistencies' section on p.19.
4. *Mostcash Plc (formerly UK Paper Plc)* v *Fluor Limited* (11 January 2002) (TCC) citing *Investor's Compensation Scheme* v *West Bromwich* Building Society (1998) 1 All ER 98.
5. French Civil Code, Article 1157.
6. *State of West Virginia* v *Ottis Ray Euman* No. 29700.
7. French Civil Code, Article 1163.
8. See 'Contra Proferentum' section on p.64.
9. French Civil Code, Article 1162.
10. *Whiteley v Chappell* (1986) LR 4 QB 147.
11. Sometimes referred to as the rule in Heydon's case (1584) 3 Co. Rep. 7a.
12. *Lister* v *Forth Dry Dock and Engineering Co. Ltd* (1989) 1 All ER 1134.
13. Provisions relating to employment rights on the transfer of an undertaking are contained in the Transfer of Undertakings (Protection of Employment) Regulations 1981 (SI 1794), as amended by the Transfer of Undertakings (Protection of Employment) (Amendment) Regulations 1987 (SI 442), the Trade Union Reform and Employment Rights Act 1993, the Collective Redundancies and Transfer of Undertakings (Protection of Employment) (Amendment) Regulations 1995 (SI 2587) and the Collective Redundancies and Transfer of Undertakings (Protection of Employment) (Amendment) Regulations 1999 (SI 1925).

Damages (including Consequential and Indirect Loss/Damage)

D

Damages act as a means of compensating an innocent party for loss or harm suffered as a result of another party's breach of contract, negligence or defamation. The loss or harm could be financial loss, physical damage to person or property or damage to reputation.

Damages can be either liquidated or un-liquidated. Liquidated damages are where a specific, usually pre-agreed, sum is requested whereas un-liquidated damages are when it is left to the courts to award what it considers to be fair and reasonable in the particular circumstances, as is the case in particular respect to personal injury and defamation cases.

The generally accepted rule says that contractual damages should be sufficient to compensate for 'such losses as may fairly and reasonably be considered as either arising naturally (i.e. according to the usual course of things) from the breach of contract'[1] or 'such (losses) as may reasonably be supposed to be in the contemplation of both parties at the time they made the contract, as the probable result of such breach'.[2]

The first of these simple rules is referred to as 'general' or 'ordinary' damages and the second rule as 'special' or 'extraordinary' damages. Everyone is taken to be aware of what loss will probably result from a breach of contract in the usual course of things (i.e. under the first rule) and anyone in breach of contract will also possess special knowledge relating to the contract works outside the ordinary course of things (i.e. under the second rule) which could give rise to additional loss possibly of a different type or to a greater extent than loss under the first rule.[3]

In order to justify an entitlement to damages for breach of contract, the injured party will have to prove that:

- the breach actually caused loss;
- the particular loss is of a type which is recognised as giving an entitlement to compensation;
- the loss is not too remote (as expressed by the two rules noted above);
- the quantification of compensatory damages is fair and reasonable under the circumstances.

The burden of proving that the breach has actually caused loss rests on the claimant and it will need to produce contemporary records in support of its claim. Generally proof will be discharged through the contractor being able to prove its case on the balance of probabilities; however, the quantification of compensatory damages must be based upon factual records and not upon theoretical calculations. Once more we stress the need to keep accurate and comprehensive contemporary records.

The purpose of a claim for compensatory damages is to return the claimant to the position it would have been in had the breach not occurred, it is not to give the claimant additional profit.[4] In a Scottish case[5] a financier broke a contract to invest

£15,000 in the business of a timber merchant and instead invested the same sum in a distillery. It was held that the timber merchant's damages were based on the loss to its business and not on the much larger profits that the financier had derived from the distillery.

The level of recovery is the loss to the victim, not the gain by the defendant. In the case of *Surrey County Council v Bredero Homes*,[6] Bredero purchased development land from the Council and covenanted to develop it in a certain way. They breached the covenant and obtained planning permission to develop the land in a different way from that to which they had covenanted with the Council. The Council sued for the amount they might have received for agreed modifications to the covenant. The remedy for breach of contract is an award of damages, and damages at Common law are intended to compensate the victim for its loss. They are not intended to transfer to the plaintiff, if it had suffered no loss, the benefit that the defendant had gained by its breach of contract. The Council had suffered no loss and therefore was only awarded nominal damages.

The measure of loss could be either based upon loss of bargain, wasted expenditure or restitution. Loss of bargain is the normal measure adopted in breach of contract claims and calculations would be based upon the actual costs incurred in putting the claimant back in the position he would have been in but for the breach. These could be the reasonable costs of completing work that a sub-contractor failed to complete or of remedying defects that a contractor failed to remedy. Where it is not reasonable to correct defects for a particular reason (for example, where the cost of remedial work is out of proportion to the value of the development) it might be that the damages would be assessed based upon the reduction in market value of the development resulting from the existence of the defect. Loss of profit would also come under this heading where the breach prevented the claimant from achieving the level of profit it would have expected to achieve but for the occurrence of the breach (for instance, where a consultant failed to apply for planning approval on a developer's site, thus diminishing its expected resale value[7]).

Wasted expenditure would apply where the claimant had incurred expenditure that was subsequently rendered futile by the defendant's breach, even if the expenditure was made before the contract was entered into.[8] If the defendant were to have any chance of disputing the claim the onus would be on the defendant to show that the wasted expenditure had been incurred through an unreasonable, unrealistic or ill-advised approach to the situation by the claimant.

The other measures of loss frequently encountered are 'Quantum Meruit'[9] and 'Restitution'[10] and each of these is discussed in separate sections of this book under their respective headings.

Consequential losses are losses that a claimant may prove over and above those that arose as a direct result of breaches of contract or negligence (i.e. consequential loss approximates to loss within the second rule in *Hadley* v *Baxendale* noted earlier in this section). These could be losses incurred by an employer as a result of a delay in completion of the works which result in it being unable to provide its customers with the product it had promised on the date it had promised, which in turn was dependent upon the contractor completing the works by that date. Consequential losses could also be incurred by the contractor, for example through loss of productivity where the claim event causes impacts in one or more trade or section of the project works.

The impact of a delay or disruption to one particular trade in one particular area may be felt immediately during or after the event or even some considerable time later. The contractor may have been required to perform otherwise unnecessary work, to alter the sequence of work, to use inefficient methods of performance, to work at inconvenient times, to perform during periods of interference or to stop work altogether as a result of the claim event. These contributory factors can be categorised as being related to:

- project conditions including:
  - overlapping of tradesmen in a particular area;
  - changes to crew size, detrimental effect on rhythm of team and productivity;
  - access restrictions to particular areas of site, necessitating double handling materials, congestion of labour and plant etc.;
  - material control and handling issues;
- learning processes, interruption of repetition and additional learning curve requirements caused by:
  - workforce reassignment resulting in disruption, loss of momentum, learning curve for each new assignment (possibly more of an employer's claim defence than a contractor's claim, depending on reasons for reassignment);
  - workforce being taken off specific work temporarily due to lack of information, drawings, approvals etc. from employer, returning at a later date and experiencing a repetition of the initial learning curve, albeit to a lesser extent – disruption, loss of momentum etc.;
  - worker turnover for reasons beyond contractor's control, resulting in excessive learning curve requirement, disruption, loss of momentum etc.;
- environmental factors:
  - work reassigned to a different season, for example sub-structure works moved from dry summer to wet winter conditions resulting in changes in work processes, shorter daylight hours, extended period etc.;
- ripple effect:
  - if certain activities are delayed, concurrent work can be expected to slow to

  keep pace;
  - if one task is delayed, subsequent tasks may also be affected;
- overtime:
  - excessive or unplanned overtime can result in reduced productivity (due to fatigue, lowering of morale, reduced supervisory effectiveness etc. also giving rise to secondary losses such as poor workmanship, increase in accidents and absenteeism);
- management:
  - where management time is focused on specific changes their attention is taken off their other duties, diluting their effectiveness and reducing time available to them to plan and coordinate both the affected and other activities not necessarily restricted to the project concerned.[11]

Quantification of consequential and indirect loss or damage claims is a complicated process and as ever relies upon good and comprehensive contemporary records. These must be specific to the affected areas of work and also to those areas unaffected, so that comparisons can be drawn as a means of justifying the impacts claimed. The greatest difficulty probably lies in linking the ripple effect on subsequent but unchanged work with the claim event.

The US Corps of Engineers have attempted to devise a formula to be used in disruption calculations that works well in their particular operating environment; however, each project is different and it is advisable to develop a company or project specific formula rather than rely upon a standard format. Developing such a formula can be a complex and time-consuming exercise but may well be cost effective when consideration is given to the subsequent litigation time that it might save. Any quantification must take due account of such factors as cause and effect relationship, allocation of responsibility and concurrency of events.

It is common in commercial contracts to specify the categories of loss that the parties will be entitled to recover in the event of breach and also to seek to exclude liability for certain other categories of loss. Contractors involved on major international contracts try to avoid responsibility to the employer for consequential and indirect loss and damage. Such a situation might arise for instance should a power plant be put out of commission due to a defect in the contractor's supply.

This could result in the employer suffering loss of revenue from its existing customers or loss of new orders due to the defect-caused forced plant outage engendering a perception amongst its potential customers of its company's unreliability in providing a continuous electricity supply. The employer might even be in breach of its electricity off-take agreements resulting in its existing customers suing it for their own loss of revenue due to their inability to manufacture and supply their own customers.

Most contractors would certainly want to be exempted from responsibility just as their employer would want to be protected.

In insurance terminology, consequential loss or damage, as opposed to direct loss or damage, is indirect loss or damage resulting from loss or damage caused by a covered peril. Take the example of a situation where loss and damage caused by typhoons are covered under a supermarket's building insurance policy. A crane working adjacent to the building is blown over in a typhoon and cuts electricity lines used to power the supermarket, resulting in frozen goods stored in the supermarket's freezers being spoilt. If the insurance policy extends coverage for consequential loss or damage then the consequential damage to the frozen goods would be a covered loss.

It is possible to take out Business Interruption insurance that provides businesses with consequential loss or damage coverage for such items as additional expenses, rental costs of alternative premises or equipment, loss of profits and commissions, etc. which are incurred as a consequence of some act or event. Such a policy would have benefited the supermarket in the above example.

Construction industry terminology would put a similar interpretation on consequential loss or damage as in the insurance case noted above. Consequential loss or damage could result from a cause or event for which the contractor was responsible, such as a defect in part of the works that it had supplied and/or constructed. Referring back to the previous example, where there is a defect in a power plant which is remedied by the contractor during the defects liability period, the contractor will, unless there is a suitable clause excluding it from any responsibility for consequential damages, be liable to the employer not only for remedying the defect but also for such compensation as the loss of power export during the period when the plant could not operate.

Where the defect was such that the power plant could not operate until it was repaired, the contractor's liability would start at the point when operation ceased and would continue until the repair process was completed and the plant was able to run again in the manner intended.

The position under the French Civil Code is stated as:

> *Even in the case where the non-performance of the agreement is due to the debtor's intentional breach, damages may include, with respect to the loss suffered by the creditor and the profit which he has been deprived of, only what is an immediate and direct consequence of the performance of the agreement.*[12]

This, it is proposed, is basically the same as the approach under Common law described above. The consequential damages noted in the last example (loss of power export and hence the loss of profit on its sale) would be directly consequential, whereas the possible loss of business suffered by the potential purchasers of that power would be indirectly consequential and not allowed.

1. 1st Rule from *Hadley* v *Baxendale* (1854) 9 Ex. 341.
2. 2nd Rule from *Hadley* v *Baxendale* (1854) 9 Ex. 341.
3. Concept of 'reasonable foreseeability' discussed in *Victoria Laundry (Windsor) v Newman Industry Ltd* (1949) 2 KB 528.
4. That is the position under Common law; under French Civil Law Article 1149: 'Damages due to a creditor are, as a rule, for the loss which he has suffered and the profit which he has been deprived of, subject to the exceptions and modifications below.' (the primary exceptions and modifications referred to include:
   - only damages which could have been foreseen at the time of the contract;
   - only damages and loss of profit resulting as an immediate and direct consequence of the non-performance.
5. *Teacher* v *Calder* (1899) 1F (HL) 39.
6. *Surrey County Council* v *Bredero Homes* TLR April 1993, CA.
7. *Obagi* v *Stanborough (Developments) Ltd* TLR 15 December 1993.
8. *CCC Films (London)* v *Impact Quadrant Films* (1985) QB 16.
9. See section on p.197.
10. See section on p.205.
11. *Europools Plc* v *Clydeside Steel Fabrications Limited* 17 January 2003 (Outer House Court of Session, Scotland) where claimant's management time spent organising and coordinating remedial work preventing him from devoting time to his other duties was deemed recoverable from the defendant.
12. French Civil Code, Article 1151.

Dayworks

When additional works are encountered during the term of a contract which are not similar in substance or in the method or timing of their execution to any item included in the contract bill of quantities, schedule of rates or tender price build-up for which an agreed unit rate or price exists, the contractor will usually attempt to negotiate a suitable rate or lump sum cost for the works with the employer. If the full extent or nature of the works cannot be identified prior to their commencement the more suitable method of claiming reimbursement could well be on a daywork basis.

In order to cover this eventuality it is advisable to include a schedule of daywork rates in an appendix to the tender bid documents covering those items most likely to warrant inclusion in a daywork claim during the currency of the contract. These daywork rates would cover labour (basic and overtime hourly rates), plant and equipment (hourly rental rates) and materials (costs per unit of measurement). In addition to a schedule of daywork rates the bid document should also specify the percentage mark-up that will be applied to each category of rate to allow for the contractor's overheads and profit.

The categories for which labour rates are included should be diverse enough to cover all trades and all divisions within those trades, including all site personnel from labourer to foreperson. Site management would all be covered by the percentage mark-up and would not be claimed for individually. Material rates should at least cover the basics of concrete, cement, sand, aggregate of different sizes, reinforcement of different grades and diameters, bricks and blocks, pipes and cables of different materials and sizes, paints, tiles, etc. Plant and equipment rates should include for various sizes of excavators, cranes, lorries, scaffolding, drills, grinders, welding equipment, pumps, compressors, generators etc. and, if appropriate, should include separate rates for contractor's own plant and equipment and for hired plant and equipment.

In the absence of daywork rates having been included in the contract documents, most countries have various bodies involved in construction, either on the trade, industry or professional side, that publish standard schedules of daywork rates for use in such situations,[1] particularly in respect of plant and equipment. Since these have a certain status and are recognised by the major players in that country's construction industry, the recognition in the contract documents of their acceptability in this respect can make the negotiation of the daywork account that much easier than it would have been had project-specific rates had to be agreed subsequent to the contract having been signed.

Labour rates are usually set by national agreements between management and staff/unions and are readily available through trade bodies. Failing this, actual rates would have to be extracted or extrapolated from independently-verified company accounts, as would rates for company-owned plant and equipment. Plant and equipment rates can, if hired as opposed to company-owned, be supported by actual invoices from suppliers, as would be the case with materials.

The employer may initiate the recording of certain works on a daywork basis where he considers this to be the most fair and equitable method of ascertaining their cost, and some contracts have clauses that empower the employer in this way. These clauses should not only empower the employer to issue daywork instructions but should also set out the timing and extent of the procedures necessary for the recording and verifying of costs expended in this way. Any daywork instruction from the employer should ideally be incorporated in a change order but should in any case be confirmed in writing.

It is also essential that the contractor obtain the employer's or the employer's representative's signature on each daywork record sheet as soon as it has been prepared, but preferably no later than the day after the work was executed, acknowledging that the items listed on the sheet as being utilised, and the hours recorded as being spent, on that particular item of work on that particular day are a true record. In this way there should be less chance for argument at a later date when the final account is being resolved as to precisely what work was done, when and by whom, although such signature does not in itself agree to any specific sum being due in payment of the related work (it is common practice for such signatures to be qualified as being on a 'for record purposes only' basis).

At the end of each week the Daywork Record Sheets for that week should be priced by the contractor, using the rates previously agreed, and invoices should be sent to the employer for approval and subsequent settlement. Any disputes as to the amounts actually claimed would be settled by the normal procedures incorporated in the contract.

In summary, the parties must take particular notice of the following in their daywork management procedures:

- contract clause empowering the employer to instruct that certain works be carried out on a daywork basis must be explicit and must cover all aspects of initiating, recording and verifying the works to be so recorded;
- layout and content of the daywork return form must be easily understandable;
- submission to employer must be prompt and complete;
- each daywork return must be based upon verifiable facts;
- subject matter of daywork must be clearly stated;
- source of instruction to carry out work on daywork must be clearly stated;
- time spent by and descriptions of labour, plant and equipment must be accurate;
- materials used must be accurate and justifiable using invoices, store dockets etc.;
- costing must be carried out using contract a schedule of rates or, if none available or applicable, using locally agreed labour and plant and equipment standard rates, invoices, independently-verified company accounts etc.

The typical daywork record sheet shown in Appendix 5[2] can be adapted to suit any particular contract.

1. In the UK this could be the Building Cost Information Service (BCIS) of the Royal Institution of Chartered Surveyors (RICS); other countries have similar institutional, industry or trade bodies providing similar information, albeit on a piecemeal basis.
2. See Appendix 5 for Typical Daywork Sheet on p.291.

Decennial Liability

Decennial liability was introduced in France in its current mandatory form in 1978 but was previously embodied in the Napoleonic Code, and maybe existed even earlier in handed-down traditions. This concept was introduced into several French-influenced international jurisdictions (including amongst others Belgium and Luxembourg in Europe and several Francophone countries of Africa) and in more recent years a number of other countries have adopted the same principle, including Australia, Egypt, Italy, the UAE, and Venezuela. Decennial liability insurance is basically a form of products and completed operations insurance that provides ten years of coverage to the owner of a construction project against necessary repair or replacement costs in the event of a complete or partial collapse.

The principle of decennial liability is similar to the statutory limitation of liability in England and Wales that gives liability limits of six years for simple contracts and ten years for deeds (contracts under seal). Decennial liability, as the phrase indicates, holds the contractor and the architect or engineer liable for a period of ten years; however, it is limited to building works as opposed to works of an engineering nature.

Under the French Civil Code the decennial liability extends to the entire framework and functions of use of the building, including such elements as the floors and roofs. The liability extends for ten years from the date of handover, or nine years after the end of the one-year 'flawless completion' period (sometimes referred to as a guarantee of perfect completion). This liability applies without the need of proof from the party who suffers the damage, unless the contractor is able to claim that the damage was caused by an event of force majeure, by the party suffering the damage or by a third party for whom it is not responsible. Construction materials manufacturers and suppliers are generally exempt from this decennial liability unless the parts that they have supplied are integral parts of the structure, such as prefabricated floor beams, staircases or the like.

Decennial liability is an insurable liability and most contracts for work covered by a jurisdiction where decennial liability exists will inevitably include a requirement for the contractor to provide decennial liability insurance (for example Egypt's Law 106 of 1976 (Construction Operations Guidance and Regulations) as amended by Law No. 2/1982 and by Law No. 30/1983).

Under the UAE Civil Code Article 880 the architect/engineer and the contractor are held to be jointly liable for a period of ten years for any total or partial collapse of the relevant construction or fixed installation, and also for any defect which threatens the stability or safety of the building, assuming the building life is expected to be ten years or more. This liability cannot be exempted by particular words being inserted into the contract between the parties.[1]

All actions have to be brought within three years of the collapse or of the discovery of the defect or damage. This in effect extends the contractor's liability to 13 years after handover. If the project owner fails to submit his claim within three years of the collapse or of the discovery of the defect or damage then it loses the right to claim. The architect/engineer's liability is limited by Article 882 of the UAE Civil Code[2] in that where it is only responsible for the plans and not for the supervision of the works, its liability is restricted to architectural/engineering design errors in the plans; all structural errors would be the contractor's sole responsibility.

The Standard Oman Building and Civil Engineering Works Conditions is an example of a standard form of contract that includes a clause requiring decennial liability cover, in this case in relation to defects in construction (not design) that lead to a failure in or collapse of the relevant structure. The ten years run in this instance from the date of issue of the maintenance certificate.

1. UAE Civil Code, Article 882.
2. Similar provisions to those in UAE Article 880 are set forth in the Abu Dhabi Building Code, Abu Dhabi Law No. 4 of 1983.

Delay Analysis

Delay analysis in respect of a construction dispute is the process in a claim or claim defence through which the contractor or employer has to go in order to be able to:

- establish lines of research and investigation;
- demonstrate the contractor's (or employer's) entitlement to claim (or to reject a claim against it or to counterclaim);
- present the claim (or claim defence, as the case may be) effectively.

The various stages described here are in general suitable for use by either the contractor or the employer. The initial research and review stage will help to ascertain whether the delay claims to be pursued are 'critical delays' or 'non-critical' delays. Critical delays are those that delay the project completion date whereas non-critical delays are those which affect progress at any given time but which do not have an effect upon the overall completion date of the project.

In addition there is a third category of delay claims that can be termed 'disruptive delay'. This would be seen in the consequential effects of changes to the method, timing or extent of the execution of the project which affect morale and productivity through congestion, out of sequence working, stop-go effects, re-assignment of resources and so on and which could give rise to acceleration measures, if these are possible under the contract, or ultimately to delays to project completion.

The next stage would be the investigation stage where all the factors relating to the basic areas of claim made known during the initial research stage are analysed. The need is to establish what were the specific causes of the delay, in what area of the work did the delay occur, when did the rate of progress start to be affected, and so on. The answers would then be saved into specific databases. The analyst would have to examine all contract records from the very beginning of the project, even going as far back as the initial negotiation stages where such records are available (unhappily not often the case).

The documents to be examined, in addition to the correspondence files, would include meeting minutes, programmes, delivery schedules, shipping records, drawing registers, weld records, schedules of approvals, day-work records, labour records, QA/QC records, method statements, notices, site photographs, payment applications and certificates, cost records, completion certificates, punch lists, warranty defects lists, insurance claims and so on.[1]

Once recorded on a database these records would then have to be analysed, summarised and put into a format that could be used to demonstrate how the particular events led to the delays. Programme comparisons would be prepared using the timing and sequence of events discovered during the research stage. Progress would be analysed not only on the project as a whole but also on specific sections of work, comparing the programme for each section with the resources expended on that section (this could show, for example, how productivity in an area not affected by a claim situation was greater than that in a similar area which was so affected). Drawing approvals would be analysed to see if they resulted in work stoppages due to lack of sufficient information to proceed. Material delivery delays, whether caused at the procurement, shipping, customs clearance or delivery stages, would also be analysed to establish responsibilities and to see what effect these delays had on the progress of the works.

After compiling databases regarding the various delay claims, the results may be shown in the form of bar charts, graphs, histograms, line graphs, charts and so on; in fact they may be represented in whatever format will make the most presentable and convincing argument when presenting the claim.

Demonstration of entitlement using the tools resulting from the detailed investigation carried out in the initial stages is the most critical element of the whole delay analysis procedure. There are several methods which can be used depending in part upon what records are available for review/utilisation and when the delay analysis procedures commence. It is advisable to think ahead and prepare all records and documentation from the very beginning of the contract negotiations with a view to having the best possible records available and procedures in place should a delay event occur later in the contract.

There are two main method groupings available for the use of contract/claim professionals, namely 'theoretical based methods' and 'actual based methods', and these general groupings each contain several methods of demonstrating entitlement from which the method most suitable for the particular project and problem concerned can be chosen.

Theoretical based methods (also called entitlement or model based methods)

Global impact method:

- the simplest method which at its most basic works on the principle that an extra week of work entitles the contractor to have an extra week added on to the contract completion date;
- not a practical method except for the smallest and most straightforward of contracts.

Net impact method:

- similar to global impact with the added refinement that it takes account of concurrent delays.

As-planned but for...:

- impacts the base line schedule with the delays the claimant believes it was responsible for;

- compares the resultant impacted completion date with the actually achieved completion date;
- holds the difference as showing that, but for the defendant's delays, the claimant could have achieved, or have been closer to achieving, the impacted completion date;
- the difference between the impacted and actual completion dates represents either the contractor's extension of time requirement or the employer's entitlement to liquidated damages, depending upon which party is claimant and which party is defendant.

As-built but for...:

- the reverse of the 'as-planned but for...' method in that it takes the as-built programme and sequentially removes the last of the claimant's impacted delays in turn, latest first until all have been removed;
- once all the claimant's excusable delays have been removed from the as-built programme (i.e. the model has been fully 'collapsed') the claimant is left with a simulated as-built programme which gives a projected completion date;
- the difference between this projected completion date and the actual completion date can be claimed as the extension of time due;
- also referred to as the 'collapsible method' or the 'simulated as-built method'.

As-planned impacted:

- starts with the original base line programme and shows the theoretical effects of impacting delay events;
- projects the end date using the original sequencing and timing of the remaining activities;
- needs an original base line schedule and also a schedule of delaying events that would have been prepared under the initial stages of the delay analysis if it had not already been prepared on an on-going basis during the currency of the contract;
- also known as the POPE (programme of possible entitlement) method and is probably the most effective of the theoretical based methods in relation to large, complex claim situations.

Actual based methods

As-planned versus as-built:

- most commonly-used method on small to medium sized projects where the claims/delays are not complex;
- as-built programme is compared with the original base line programme – these are probably in bar chart format on the type of project for which this method is best suited;
- it is relatively simple then to establish which events caused which particular

period of delay and to visualise the concurrency aspects of these delays and their criticality;

- the main advantage of this method is its simplicity and the fact that it gives a visual interpretation of events in a simple format;
- not suitable for complex projects where there are large numbers of activities and competing delay events to be considered, possibly in several different work areas at different times.

Windows/snapshot or update method:

- advancement of the 'as-planned versus as-built' method using basically the same procedures but utilising critical path network analysis in place of simple bar charts;
- this added refinement makes it more suitable for more complex projects and delay;
- overall project duration is divided into 'window' periods that should match the reporting periods used on that project (most probably monthly);
- actual work carried out in a window as shown in the as-built programme is imposed upon the as-planned programme, taking account of any programme revisions made during that period;
- overall end date is calculated taking account of the information gleaned from the current window in relation to that resulting from similar analysis of the previous window;
- it is therefore possible to identify the delay that occurred within each window period and further research will then establish the causative event;
- this method takes account of such items as mitigation, criticality, culpability and concurrency, gives a general feeling of openness and provides a more complete overview of the claim.

Impact/update method:

- an improvement on the 'windows' method, based upon similar principles but bringing causation into the window analysis at an earlier point in time but:
  - relies heavily on good as-built records being kept; and
  - is time consuming to prepare;
  - it is therefore more expensive than the other methods described; but
  - the advantages of its use on complex projects considerably outweigh the disadvantages;
- delay schedule is first prepared for the window period;
- delays are impacted into the window as-planned schedule and time analysed;
- result gives a projected completion date that indicates the extension of time entitlement;
- contractor's culpable[2] delays during that period are then impacted on the model

and the resulting project completion date thus includes both culpable and non-culpable[3] delays;

- this analysis is followed by an analysis of the as-planned model with actual progress imposed upon it, resulting in a completion date that represents the actual delay caused during that window period;
- takes account of excusable delays, compensable delays, culpable delays and mitigation actions taken by the contractor;
- analyst can then from these two separate analyses resolve issues of concurrency dominance, mitigation etc.

The windows and impact update methods are the preferred methods on complex projects, not only because they are better suited to the complex claims that can occur on the larger projects but also because they are more transparent, and practical in their application. The fact that they are based upon a critical path analysis gives them added credibility, as does the fact that they take account of actual progress and have a level of objectivity that the other methods described, with a few noted exceptions, lack. They take account of the events at the time that they occurred and consider the likely entitlement at the end of each window period, whether that be each month, at each milestone or whatever depending what window period has been chosen for the analysis.

Despite the windows and impact update methods being the preferred methods they cannot always be used. There might be time constraints, which need a more compact method to be used, or budget restraints that require a less expensive route to be taken. There might not be sufficient as-built records available for the preferred option to be adopted. If at all possible it is helpful to try two different methodologies and compare the results in order to test the strength of the chosen methodology. It is advisable also to try making minor adjustments to inputs to see what effect they have on the projected completion dates and whether these effects are reasonable, justifying the methodology, or unreasonable, putting its suitability into question.

1. See full list in 'Claim Review and Preparation' section, Step 1, p.47.
2. Contractor's culpable delays – those delays for which the contractor is to blame.
3. Contractor's non-culpable delays – those delays for which the contractor is *not* to blame.

Depositions

A 'deposition' is a legal term for 'asking questions'. Basically, the use of depositions gives each party the opportunity to prepare for trial by asking questions of the other in order to gain vital information. Parties to the suit, witnesses and experts may all be deposed. The deposition process is common in the US for taking evidence from witnesses prior to trial and is also a useful means of obtaining evidence from witnesses who would be unable or unavailable to attend the trial to give evidence in person.

This latter case is called a deposition 'de bene esse'[1] and is not to be given in evidence without the consent of the opposite party, unless the witness is dead, beyond the jurisdiction, sick or infirm. This is rarely applied in domestic construction disputes in the England and Wales, or indeed in many European countries, but is a useful means of obtaining evidence in disputes of an international nature, as long as it does not contravene any local laws or regulations. The local situation must be checked thoroughly with lawyers familiar with the constraints, if any, of the local legal system before proceeding in this manner.

A deposition presents a preview of what the deponent will say at trial. If the deposition goes well, this is a unique opportunity to hear the deponent's testimony weeks or months before the hearing or trial and this will give an insight into the opponent's arguments, specific demands, strategies and specific allegations. While a deponent can change his testimony at trial this will obviously put a big question mark over his credibility with the court. A deposition testimony can, if managed effectively by the other side, be used at trial to expose contradictions in their opponent's case.

The other principal purpose of a deposition is to obtain information that the opposing attorney needs in order to further his case to the best advantage. Each case, to be successful, must be founded on facts, and a deposition is a sound way of obtaining information from an opponent upon which the case can be built or at least improved. The information obtained may be used offensively or defensively but in either case it can represent the starting point for the search for further information expanding on the relevant information disclosed at the deposition.

There are three main types of deposition witnesses:

- witness of fact;
- expert witness;
- witness called under Federal Rule 30(6)(b) (applies in US only).

A witness of fact is sometimes referred to as a 'lay' witness. They are called to testify on facts that they know to be the case and in respect of events which they actually witnessed relevant to the case. They would probably be employees of the opponent company or sub-contractors, suppliers, consultants and so on who were involved with the project at some time relevant to the case.

An expert witness gives the tribunal, or court, his 'expert' opinions regarding a specific aspect of the case in relation to which he has expert knowledge. An example that I recently experienced was where a dispute arose over the quality of welding on a project in the US under ASME[2] codes. In this case we were able to call as an expert a member of the ASME board who provided valuable testimony based upon his expert knowledge of the ASME codes. This was one case where the expert's particular knowledge matched perfectly with the main issue in dispute, i.e. the true intention of a particular ASME code. The expert was not just a highly qualified engineer but also a member of the ASME board, which gave his testimony additional weight. This is the sort of particular knowledge to look for when selecting an expert.

Federal Rule 30(b)(6) only applies in the US. If a party requires a deposition on a particular topic but is unsure who to depose, it would issue a deposition request under Federal Rule 30(6)(b)[3] naming as the deponent a public or private corporation or a partnership or association or governmental agency, describing the matters on which examination is requested, and the organisation so named would then designate the person best suited in the organisation to provide that deposition.

During the deposition, attorneys ask questions of the deponents in the presence of a court stenographer, in England and Wales in the presence of a magistrate or other judicial authority. The evidence is given under oath and is taken down by the stenographer and then transcribed. The witness has the right to read, correct and sign the transcript although this right can be waived. The transcript is thereafter presented as evidence to the tribunal or court. The stenographer may be asked to read back questions and their answers for the benefit of the attorneys and is responsible for marking and listing any documents put in as exhibits at the deposition. Telephone or video links are permitted in England and Wales.

The total time necessary for depositions varies widely depending on the number and complexity of the issues in question but in a major international construction dispute it would be hoped that the timescale could be limited to three or four months at the most.

Attorneys set their strategies for the deposition depending upon their reading of the deponent and of the case in general. The tenor can be friendly, informal and unstructured or formal, regimented and aggressive. Most attorneys would avoid trying to embarrass or intimidate a witness but on rare occasions this might be considered the most potentially effective strategy to pursue. Where the deposition is recorded on video, such an attitude on the part of the attorney would not help the attorney's case. In general a softer, gentler approach is more likely to bring out the best result from, in particular, a lay witness.

Since the primary objective of a deposition is to preview the other side's case and obtain information, the task of a deponent during their deposition is to give away as little information as possible to the other side. This refers not only to the basic information requested but also to the tone, style, and manner of their presentation that might give the opposition attorney an insight into the strategy that has been decided upon for the hearing.

It is a good idea for a party's attorney to coach its deponents before they actually give their deposition. I have used the following 20 'rules' as a basis for controlling depositions for many years and I have found that they contribute considerably to the quality and effectiveness of the performances of deponents at the deposition. I direct the deponent, particularly if they are lay witnesses, to:

- always tell the truth; this is the most important rule which must never be broken or compromised in any way;
- think before answering the question, not only to ensure that the question is fully understood but also in order to have time to recall all the facts necessary for a complete answer and to consider the implications of the various ways in which the answer could be phrased;
- just answer the question and be succinct; once you feel that you have answered the question, you should stop talking; stick to what you know and do not embellish the facts in any way;
- do not volunteer information not asked for;
- do not make argument;
- do not be led; if the opposing attorney introduces a statement or asks a question that assumes something that is untrue, interject courteously with a correction;
- your answer should be as accurate as possible without any exaggeration; do not make claims that cannot be backed up;
- do not guess or speculate; qualify answers with 'as far as I can recall' or 'I believe' as appropriate if not totally sure of the accuracy of the response;
- if you don't know the answer you should say so, you cannot be expected to know everything;
- if you think a particular document might jog your memory you should ask to see it;
- do not be intimidated by the demeanour of the opposing attorney;
- keep cool, do not be rattled by the opposing attorney's sarcasm or plain, blunt rudeness;
- take time, do not rush to answer, thinking time is without prejudice or penalty and should be used freely; a pause before answering gives the deponent's attorney time to object to the question;
- take a break if needed, do not let your concentration wander;
- if presented with any documents, read them carefully before responding;
- do not promise the opposing attorney to do something or to be available for further questioning without first discussing that with your own attorney;
- ask to speak to your own attorney if faced with a crisis or dilemma, do not give your attorney any surprises during your deposition, and let your attorney formulate a position on the matter before responding;

- stay on guard and be alert at all times;
- maintain a professional demeanour, ensure your appearance is suitable for the occasion, that your dress is clean and neat; and treat all persons involved with courtesy and respect;
- do not fill in silences by talking, you are just there to answer questions posed by the opposing attorney; silence is okay.

Expert witness depositions play an ever-increasing part in dispute resolution processes. In the US, Federal Rules of Civil Procedure allow for the introduction of expert witnesses into the dispute procedure.[4] Experts should not need the intensive coaching just described for lay witnesses, although it would do no harm to remind them of the points noted. They should, however, be coached in respect of a few points where experts frequently fall down. The following seven points are suggested as being of particular relevance and importance in this respect:

- witnesses have a right to read and sign their deposition; opposing attorneys often agree on general stipulations at the outset and one that frequently arises is the stipulation waiving the rights to read and sign depositions. In the interests of accuracy experts should object to this stipulation;
- experts need breaks in the deposition-taking process as much, if not more, than lay witnesses do and should be encouraged to take frequent breaks in order to maintain their concentration and collect their thoughts;
- experts should have in-depth discussions with their attorney before the hearing in order to discuss strategies, to find out how the hearing is going so far, to review possible questions that might be asked and to establish any legal standards pertinent to the expert's testimony;
- experts should ensure that their CVs are accurate and that there is nothing there of a questionable nature that could be used to destroy the expert witness's credibility;
- experts should not tamper with records; if it can be shown by the opposing attorney that the expert witnesses have tampered in any way with their files, by removing a particular letter or document for example, then this will again seriously damage their credibility.

When giving opinions, experts should have previously thought out the possible questions which these opinions might raise and should, accordingly, be able to defend them effectively; they should be able to demonstrate the facts and assumptions upon which their opinions are based, the method used in arriving at the opinion stated, when that opinion was first formed and the documents which assisted such formation, the degree of flexibility within that opinion and how the opinion relates to answers previously given.

Experts tend to be persons who have a high regard for the knowledge that they possess which is, for the most part, why they have been called as witnesses. This results in a tendency to over talk. As noted in the advice for lay witnesses, volunteering information is to be frowned upon and so expert witnesses have to be particularly careful not to let their own egos run away with themselves.

When taking depositions in one country for use in another it is emphasised that great care must be taken to ensure that the act of deposing or the manner in which it is carried out is not contrary to the laws of that country or to any treaties between the two countries.[5] Notwithstanding any treaties or conventions entered into by the countries concerned, consideration has always to be given to local laws and regulations and to any other specific agreements between the countries. For example:

- the Government of Brazil asserts that, under Brazilian constitutional law, only Brazilian judicial authorities are competent to perform acts of a judicial nature in Brazil. Brazil has advised it would deem taking depositions in Brazil by foreign persons to be a violation of Brazil's judicial sovereignty. The Brazilian prohibition on taking depositions by foreign persons extends to telephone or video teleconference depositions;
- in Japan all depositions by US[6] attorneys for use in the US must occur at the US Embassy or one of its Consulates, and they must be taken in the English language and be overseen by a US Consular officer. In addition, the Government of Japan does not permit video or telephonic depositions.[7]

1. Literally: 'as well done (as possible)' – i.e. the taking of provisional, temporary action in anticipation of a future occasion. In this context, to take evidence for future use while it is available.
2. Acronym for 'American Society of Mechanical Engineers' – amongst other things, ASME sets internationally-recognised industrial and manufacturing codes and standards.
3. US Federal Rules of Civil Procedure (2003) Rule 30(b)(6) states 'A party may in the party's notice and in a subpoena name as the deponent a public or private corporation or a partnership or association or governmental agency and describe with reasonable particularity the matters on which examination is requested. In that event, the organisation so named shall designate one or more officers, directors, or managing agents, or other persons who consent to testify on its behalf, and may set forth, for each person designated, the matters on which the person will testify. A subpoena shall advise a non-party organisation of its duty to make such a designation. The persons so designated shall testify as to matters known or reasonably available to the organisation. This sub-division (b)(6) does not preclude taking a deposition by any other procedure authorised in these rules.'
4. US Federal Rules of Civil Procedure (2003) state in Rule 26(a)(4): 'A party may depose any person who has been identified as an expert whose opinions may be presented at trial. If a report from the expert is required under sub-division (a)(2)(B), the deposition shall not be conducted until after the report is provided.'
5. Refer to the 'Hague Convention on the Service Abroad of Judicial and Extra-Judicial Documents in Civil and Commercial Matters' and the 'Vienna Convention on Consular Relations' for more detailed information in this respect.
6. Article 17 of the US–Japan Consular Convention authorises US consular officers to take depositions in Japan, 'on behalf of the courts or other judicial tribunals or authorities of the sending state (United States), voluntarily given, in accordance with the laws of the sending state (US) and in a manner not inconsistent with the laws of the receiving state (Japan).'

7. Check with the Commercial Section of your country's embassy in Japan for similar restrictions before undertaking a deposition.

Discovery

Discovery can be defined as 'the process whereby the parties to a lawsuit exchange and disclose to each other all documents and information in their possession, custody or power relating to matters in question in that lawsuit'. Discovery includes both disclosure[1] and inspection.

'Documents' in this context can mean:

- any document in writing, including agreements, correspondence, minutes of meetings, reports, entries in diaries, notes on the back of envelopes or on odd pieces of paper, etc.;
- drawings, photographs, graphs, diagrams, schedules, programmes etc.;
- tapes, videotapes, disks and computerised information (see later discussion on computerised information);
- generally, anything on which information of any description is recorded.

A document has been defined[2] in England and Wales as 'anything in which information of any description is recorded' and so almost every form of information transmission, retention, storage, presentation or display is capable of being construed as and existing as a document.

The word 'possession' in the definition of 'discovery' means the right to possession of a document and it does not require, though clearly includes, actual physical possession.

The word 'custody' in the definition of 'discovery' means the actual handling of a document regardless of the right to possession (for example, if a party was holding a document belonging to someone else, possibly a sub-contractor, which was relevant to the action in question, that party would have to disclose the document).

The word 'power' in the definition of 'discovery' means the enforceable right to inspect the document or to obtain possession or control from the person who has physical possession of it without having to obtain the consent of anyone else.

Defences to discovery are 'privilege' and 'legal professional privilege'. Privilege is the entitlement to refuse disclosure of a document the existence of which is discoverable. Legal professional privilege would apply to any communication between a client and lawyer in the course of their professional legal relationship, which communication was clearly written for the purpose of obtaining or rendering legal advice, provided such advice was given. Communications dealing with legal assistance are discoverable, as opposed to communications dealing with legal advice that are not discoverable since when a client seeks legal advice there is a possibility of litigation being involved at some stage and legal professional privilege would then kick in as a defence.

Discovery is peculiar to common law jurisdictions and even within those various jurisdictions there are differing rules and methods of application in effect, the majority of which are more restrictive than those of England and Wales. Some

Civil law jurisdictions, such as Switzerland, have a few limited exceptions where discovery is covered by statute. Particular rules and regulations, laws and statutes are subject to change and readers are required to familiarise themselves with those rules, regulations, laws and statutes in the particular country concerned at the time any action is contemplated.

The two principal types of discovery are:

- voluntary; or
- court ordered.

In England and Wales it is possible[3] for a party to ask the court to order 'specific disclosure', a process similar to court-managed discovery in the US.[4]

Parties may obtain discovery by one or more of the following methods:

- depositions[5] upon oral examination or written questions;
- written interrogatories;[6]
- production of documents or things or permission to enter upon land or other property for inspection and other purposes;
- physical and mental examinations; and
- requests for admission.

Interrogatories are useful in complicated cases where the expense of an examination for discovery[7] would be high.

In England and Wales existing Civil Procedure Rules (CPR) require the disclosure of documents on which the party relies or which 'adversely affect his own case'.[8] This is usually after the close of pleadings in an action begun by writ and relates to documents that are or have been in the party's possession, custody or power relating to the matters in question in the action. Discovery would initially be in the form of a list of such documents.

The lawyers for each party are required to perform a 'reasonable search'[9] and are provided with four considerations for determining what is 'reasonable', including 'the ease and expense of retrieval'. The resulting disclosure must identify any category or class of documents for which the lawyer has determined a search would be 'unreasonable'. The opposing party has the right[10] to inspect and make copies of any disclosed document, unless the disclosing party states that the document is no longer within its control, the document is privileged or disclosure would be disproportionate to the issues in the case.

In the US the process is controlled by the Federal Rules of Civil Procedure (FRCP) and by State rules patterned after these rules. Their objective is to provide safeguards against discovery abuse. The lawyers on both sides must, within approximately 90 days of the start of litigation, disclose 'all documents, data compilations, and tangible things'[11] in their possession, without a request from the other side. The use of the term 'data compilations' by those responsible for writing the latest FRCP suggests that anticipated the increasing use of computer-based disclosure. The scope of FRCP disclosure is narrowed, however, to only those 'documents, data compilations, and tangible things' that the lawyer intends to use in the litigation, not necessarily all those that are relevant. Under the CPR of England and Wales the scope is broader in that the lawyer is required to disclose documents that it may not plan to use and even documents that it may in fact view as being detrimental to its case.[12]

In the US every request for discovery or response to a request for discovery must now be signed by the attorney concerned to certify that the request is consistent with the FRCP and that it is not made for improper purposes or is unreasonable or unduly burdensome or expensive. In the US there is no automatic duty to render documents or to produce witnesses but objections may be made and privileges invoked similarly to England and Wales. For example, a party may request a protective order from the court that has power to order that confidential trade secrets not be disclosed.

Generally, the party who has delivered a list of documents shall allow inspection and copying of all those items that have not been objected to. The list must include a notice stating a place where the documents may be inspected and copied. Copies can be made at the designated place on payment of the costs of reproduction, delivery and so on.

The court may, at any time on the application of a party, require any other party to deliver an affidavit[13] stating whether a certain document has been in the other party's possession or control and, if not then in that party's possession or control, when the party parted with it and what has become of it subsequently.

If a claim for some type of privilege, which would shield the document from being given to the other party, is to be made then such claim must be made in the list, together with a statement of the grounds of the privilege claimed. The court may inspect the document for the purpose of deciding the validity of the claim.

In the US, where a document is in the possession or control of a person who is not a party, the court, upon service of a notice to the person and to all other parties, may order production and inspection of the document or preparation of a certified copy that may be used instead of the original. Despite the US being a Common law jurisdiction similar to England and Wales, this is only likely to happen in England and Wales in exceptional cases.

Another distinction between discovery in the US and discovery in England and Wales is that in the US discovery has been allowed to take many forms, including on-site inspections of workplaces and depositions of potential witnesses. The scope of discovery has included previously unrecorded telephone conversations, exchanges in meetings, and even personal thoughts. In England and Wales, however, discovery is limited to documents, which has the effect of narrowing the scope of disclosure somewhat in comparison to discovery in the US.

English courts do not, except in a limited class of cases, allow discovery against non-parties to the action (the purpose of the limited exception being chiefly to enable a person injured in an accident to obtain access to his own hospital records if he wishes to sue). As far as the parties to the litigation are concerned, generally speaking, they must disclose:

- documents which are admissible in evidence or disprove any of the disputed facts (the facts in issue);
- any document which, though not itself evidence, 'contains information which may (not must) either directly or indirectly enable the party [applying for discovery] either to advance his own case or to damage that of his adversary or if it is a document that may fairly lead him to a train of inquiry which may have either of these two consequences.'[14]

The disclosure, or the production for inspection, of a document is not in itself an admission of its admissibility.

Where, after a list of documents has been delivered, it comes to the attention of the party delivering it that the list was inaccurate or incomplete, or an important document comes into the party's control after the delivery of the list, the party must immediately deliver a supplementary list specifying the inaccuracy or the document. Where a party fails to make discovery of a document, or fails to produce a document for inspection or copying, the party may not put that document in evidence in the proceedings or use it for the purpose of examination or cross-examination unless they have the express permission of the court to do so.

Following on from the quotation above from the *Peruvian Guano* case, it is clear that parties may obtain discovery regarding any matter, other than privileged matter, which is relevant to the subject matter involved in the action concerned, including the existence, description, nature, custody, condition and location of any books, documents or other tangible things and the identity and location of any person having knowledge of any discoverable matter. If the information sought appears reasonably calculated to lead to the discovery of admissible evidence then the party upon whom the discovery notice has been issued would not be able to object to production of the information on the grounds that the information sought will be inadmissible at the trial.

An interesting recent development has come about in Hong Kong where a loophole concerning discovery of transcripts has been closed. In these technology-conscious days it has become possible for nearly instantaneous transcripts of the proceedings to be made available to the parties on their laptops either in the courtroom or in their offices, providing a very convenient referencing and monitoring tool.

As this is an expensive exercise, the party intending to take advantage of such a service frequently asks the other party whether they would like to join in the use of this service and to share the cost of so doing. It has been common for the expense of this service to force the other party to decline such an offer, in which case they are left at a slight disadvantage. To overcome this perceived disadvantage the declining party has taken in some instances to making a request to the courts for discovery of the transcript under Order 24 RHC, thus obtaining the same service at a fraction of the cost. This was obviously unfair and to prevent this practice getting out of hand a new rule 11B has been added to Order 24 RHC in Hong Kong, resulting in transcripts no longer being discoverable.

The current issue that is generating much discussion in the US (not so much in England and Wales as yet) is the issue of how far computer-based information should be permitted in discovery. Surprisingly, paper discovery is still the major form of discovery and it will be interesting to see how the rules are changed to accommodate new technology.

1. The term 'discovery' is generally used to cover both discovery and inspection. Discovery is strictly speaking the process of disclosing that a document exists and inspection is the process of producing the relevant documents to the party who has given discovery. Discovery and inspection/production are sometimes called collectively 'Disclosure', for example in Queensland Supreme Court Rules, Australia.
2. As defined in Rule 31.4 of the Civil Procedure Rules of England and Wales (CPR).
3. As defined in Rule 31.12 of the Civil Procedure Rules of England and Wales (CPR).
4. US Federal Rules of Civil Procedure – Rule FRCP 26(b)(1).
5. 'Depositions' are sworn testimonies from witnesses, usually involving the lawyer for the party seeking the deposition asking questions with the other party's lawyer in attendance and participating as deemed necessary (see previous Section on p.91).
6. 'Interrogatories' are written questions sent from one side to another with the objective of receiving written answers to questions of fact or explanations of legal contentions made by the other party, usually made on affidavit (see following Section on p.152).
7. An 'examination for discovery' is an oral examination on oath by one party in an action of another party.
8. As defined in Rule 31.6 of the Civil Procedure Rules of England and Wales (CPR).
9. CPR Rule 31.7.
10. CPR Rule 31.3.
11. FRCP 26.
12. See quotation later from *Compagne Financiere du Pacifique* v *Peruvian Guano Company* (1882) 11 QBD at p.100.
13. An 'affidavit' (from the Latin meaning 'he has declared upon oath') is a written statement sworn or affirmed before some person who has authority to administer an oath or affirmation. Meanings in the US and in England and Wales are similar, although in the US an affidavit is less likely to be substituted for oral evidence than would be the case in England and Wales.
14. *Compagne Financiere du Pacifique* v *Peruvian Guano Company* (1882) 11 QBD at p.100.

Document Control

Frequent reference is made in different sections of this book to effective document control, in particular in relation to the review, preparation and defence of claims.[1] Various documents have been suggested as being essential to the claim review, preparation and verification stages, without which it would be difficult to prepare a claim, to justify it to the satisfaction of those deciding its merit or to review it in detail from a claim defence standpoint.

This section will try to identify a more comprehensive listing of documents which should in an ideal world be maintained in relation to every project and which should be readily available at any time. Some will be more readily available to the employer, some to the contractor and some to the sub-contractor but most would be common to, or available upon request to, any party preparing or reviewing a claim.

Bid documents:

- client's brief/project requirements;
- invitation to bid;
- instructions to bidders;
- addenda and supplementary information;
- terms and conditions of contract, plans and specifications;
- minutes of pre-bid meetings;
- question and answer sheets from and to each bidder;
- risk analysis;
- site inspection report;
- environmental impact report;
- estimate breakdown;
- labour and equipment schedules;
- material and equipment vendor quotations;
- sub-contractor quotations;
- programme drafts;
- discussion notes – internal and with prospective client, including telephone logs;
- bid submission;
- bid competitor analysis.

Contract (or sub-contract) documents:

- negotiation meeting minutes and discussion notes;
- contract plans and specifications;
- general and special conditions of contract;
- contract programme/CPN;
- performance and other relevant bonds;
- original of executed contract documents;
- sub-contract and material/equipment purchase orders;
- building permits;
- import licences.

General correspondence and data:

- incoming and outgoing correspondence (letters, memos, e-mails, faxes etc.);
- minutes of meetings (segregated into project site meetings, internal company/consortium meetings, quarterly progress meetings, sub-contractor meetings, scheduling meetings, QA/QC meetings, commissioning meetings etc. as appropriate);
- Requests for Information (RFI) submissions;
- change orders;
- notices (interruptions, delays, changes, contractually required time related notices etc.);
- internal memos and e-mails;
- transmittals;
- progress reports;
- site progress photographs;
- site, management and supervisors diaries;
- trade specialists' reports;
- accident reports;
- material and equipment shipping information (airway bills, bills of lading etc.) and customs clearances;
- material receiving reports complete with OS & D^2 reports, photographs of shipping and packing issues, damage etc.;
- weather logs;
- daily records of activities;
- soil reports and site survey data;
- piling records;
- concrete pour logs.

Cost records:

- project cost breakdown and budget for each section of work;
- cost control reports;
- change order estimates;
- dayworks;
- back-charge records and summaries;
- labour, plant and equipment usage charts;
- material purchase orders, delivery receipts and payment receipts;
- sub-contracts price breakdowns, sub-contractor payment applications, payment receipts and status reports;
- payroll records;
- contractor's payment applications to employer;
- employer's payment records;
- finance records (bank and loan statements and reports etc.);
- cash flow projections and budgetary control reports.

Quality control:

- quality control programme;
- method statements;
- shop drawings;
- record drawings;
- copies of all relevant Standards to be applied, including BS, DIN, JIS, ASME etc.;
- off-site material and equipment test reports;
- on-site material and equipment test reports;
- inspection reports;
- QA/QC reports;
- environmental control plan;
- test reports and sign-offs;
- punch lists;
- certificates of substantial completion, partial completion, final acceptance, etc. as appropriate;
- warranty defects control records.

Schedule control:

- original contract CPN schedules;
- updated contract CPN schedules (each update);
- bar charts;
- time-forecasting charts;
- work breakdown activity charts;
- project management plan.

The above list is by no means exhaustive and the record topics will depend to a large degree upon the particular employer's, contractor's or sub-contractor's modus operandi and the size, nature and location of the project. However, these should be looked upon as a minimum requirement. If they can be coded and linked to a computerised document control system then this will be a distinct advantage, particularly at the end of the project when many of the original staff have left and it is left to the contract close-out team to try to find sufficient factual evidence to identify, prepare, submit, support or defend claims. Wherever possible summaries should be prepared and updated regularly, logging which document has been recorded where and in which file it is located.

1. See 'Claim Prompt List' on p.38 and 'Claim Review and Preparation' on p.47.
2. Over, Short and Damage reports – reports prepared on examining incoming material shipments – See Appendix 6 on p.293 for suggested OS & D Report format.

Duty to Warn

In recent years there has been a noticeable change to the traditional view under the Common law of England and Wales that if a contractor was executing its works in accordance with the contract it had no separate obligation to comment on either the suitability or danger of the design or specification. It is now accepted that where a contractor or sub-contractor is aware that there are design defects in the structure that it is constructing, it has a duty to warn the employer about those defects.

The crucial wording here relates to the contractor actually being aware of the likelihood of danger; it is not enough that a competent contractor 'should' have been so aware, although this interpretation might have force under certain specific circumstances. In one particular case[1] a contractor was constructing temporary works which were obviously dangerous and which the sub-contractor knew to be dangerous. The sub-contractor failed to warn the contractor about the dangers which it foresaw and for which it was held liable, although the extent of its liability was reduced to take account of the contractor's own negligence in not seeing the dangers for itself.

In general there is still a long way to go before more definitive guidance is available. The decision in the case noted above avoided making general rules and rather suggested that each case had to be viewed on its merits, with several factors affecting the final decision including the terms of the contract, the experience and level of expertise of the contractor, the reliance placed by the employer upon the contractor's specialist knowledge and the relationship between the contractor and the employer.

Courts have held on several occasions that a contractor is liable if it fails to warn the employer of a danger that is obvious[2] or so obvious that the contractor should have recognised it as such[3] and/or when it is judged that the danger should have been deduced by the contractor from the plans it was presented with and which the construction of the works was required to comply with.

The obligation does not only relate to the relationship between contractor and employer but also affects the other parties involved in a contract equally, including the sub-contractors, the architect, project manager, owner's representative, engineer etc. It has been held[4] that an implied term in the contract[5] required the contractor to inform the architect or engineer of defects which it believed existed in the design or as soon as it came to believe that they existed[6] but that this did not mean that the contractor was required to actively search the drawings, specifications etc. for possible defects.

It has also been held that an architect or engineer has a duty to warn its clients if it intends using new materials or techniques as yet untried, and to obtain its client's approval of their use. The English Court of Appeal emphasised the point that project managers may be in breach of their duties if they do not appropriately advise their clients about the risks attached to construction materials. The fact that the client is cost-conscious does not diminish the project manager's duty to warn against using inappropriate materials, the only basis for exoneration for not warning its client being if it is clear that the client would have ignored the advice if properly given.[7]

The extent of the contractor's liability may be diminished by the contributory negligence of the employer and/or its professional advisers. This happened both in the *Plant Construction* v *Clive Adams and JMH Services* temporary works case noted above where the judge actually reduced damages awarded to the main contractor by 80 per cent because of the contractor's own contributory negligence in not recognising the potential danger itself, and also in the *Lindenberg* v *Canning* case where the builder was held 25 per cent responsible and the surveyor 75 per cent responsible.

Sub-contractors are no less open to liability than main contractors. They have a definite liability in respect of their own work whereas their liability in relation to the contractor's work is limited by the extent of their knowledge of the main contractor's intentions in that respect. Where the main contractor could adopt several methods of construction in relation to work which the sub-contractor is or has been associated with, unless the sub-contractor knows precisely which method the contractor intends to adopt it cannot be held liable for not informing the contractor of potential dangers in one particular method of construction.[8]

In the absence of more specific guidelines by the courts it has been recommended[9] that the courts should ask the following questions when reviewing a dispute concerning a duty to warn, and these recommendations are something that contractors and sub-contractors should bear in mind throughout the execution of the project:

- should the ordinarily competent contractor have been aware of the existence or probable existence of the defects or dangers in question?
- if so, should the ordinarily competent contractor have given a clear warning to its client (or indeed anyone else) (for example, any other responsible party such as the architect or engineer)?

In the case of architects or engineers using new materials or procedures, the architect or engineer, having specialised knowledge in a particular field that in all probability influenced the client when making its initial appointment, has a particular responsibility to inform the client and to obtain its permission before using such materials or procedures. The question to be asked by the court in such a case would be were there any mitigating circumstances that precluded the need for such action on the part of the architect or engineer? (In a Design & Build, EPC or similar contract where the contractor is responsible for the design, the contractor, not the architect or engineer, would have such responsibility towards the employer.)

It is recommended that the following points be taken into consideration when implementing another party's design:

- ensure that the result which the employer is seeking can be achieved with the design as provided;
- liability for loss resulting from defects in the design should be excluded in a written contract;

- defects should be raised immediately with the employer as soon as they become apparent – whether it is before or during construction;
- notice of any defects should be recorded in a letter to the employer as proof that the matter has been brought to the employer's attention;
- a written indemnity should be sought from the employer if the employer wishes to proceed with what is believed to be a defective design;
- alterations to the design should not be made to remedy defects unless this has been discussed with and approval obtained from the employer. Failure to do so may result in an inability to claim for extras required to remedy the defective design. Alternatively, the written contract should provide the contractor with the right to make any necessary changes to the design to remedy any defects. Such a right should, however, be exercised with caution.

A contract may define the contractor's responsibilities under the contract such that it excludes liability for defects in design, although such a clause is unlikely to be effective where the contractor has 'actual' knowledge of a defect and fails to warn the employer. In the case of *University of Glasgow* v *Whitfield* it was stressed that the contractor knew that the employer placed reliance on it in the matter of design, establishing a special relationship between them and conferring a duty on the contractor to warn the employer greater than its duty to warn the architect, engineer or designer.[10]

Despite the court rulings noted above there is still much room for uncertainty as regards a party's liability in respect of a duty to warn. Questions that remain open for discussion in each particular case include:

- What is the extent of the duty of a party in an individual case?
- When does the party at risk cease being under such duty?
- What steps can the party at risk take to protect itself at tender stage in the form of insurance, price determinations or other risk management procedure.

The Common law position applied by courts in England and Wales has gained support in other countries whose Common law systems are based to some extent on the English system, notably Canada (except Quebec where law is based upon the French Civil Code), New Zealand and Australia. The Civil Code position is not quite so clear, although it would seem to provide similar results. Codified rules are often clearer to apply than Common law precedents since the latter are often subject to interpretation as opposed to the strict guidelines to be followed under a codified system. In the US, different states tend to approach the subject differently. In Louisiana, for instance, the Louisiana Design Sufficiency Law provides the contractor with immunity for defects in the work and states that:

> *No contractor, including but not limited to a residential building contractor as defined in RS 37:2150.1(9), shall be liable for destruction or deterioration of or defects in any work constructed, or under construction, by him if he constructed, or is constructing,*

the work according to plans or specifications furnished to him which he did not make or cause to be made and if the destruction, deterioration, or defect was due to any fault or insufficiency of the plans or specifications. This provision shall apply regardless of whether the destruction, deterioration, or defect occurs or becomes evident prior to or after delivery of the work to the owner or prior to or after acceptance of the work by the owner. The provisions of this Section shall not be subject to waiver by the contractor.

The basis for this reasoning is that the contractor is not a guarantor of the plans and specifications drawn up by the employer's architect, surveyor or engineer. Despite this legislation, however, the US courts have held that the contractor is not relieved from the duty to use ordinary care and skill in the execution of its work, the duty to refrain from creating hazardous conditions and the duty to warn of any hazardous conditions that it discovers, likening this with that which the ordinary ethical person should disclose under such circumstances.[11] Where a US contractor became aware of drainage problems near a housing project it was constructing and failed to inform the homebuyer of the hazard, it was held liable 'because it had breached the duty to disclose by failing to warn the plaintiff (homeowner) of the drainage problem and risk of flooding'.[12]

In order not to fall foul of the contract and the obligations of the parties it is advisable to ask oneself the same questions recommended above at the time that any relevant issue first raises its head and, if a warning is deemed to be required, to give due warning in writing. It is considered unlikely that mere mention of the defects or dangers in a telephone call or informal discussion would suffice.

1. *Plant Construction Plc* v *Clive Adams Associates and JMH Construction Services Ltd* (CA, unreported, 20 December 1999 and TCC 31 March 2000) – contractor was constructing temporary works which were obviously dangerous and which the sub-contractor knew to be dangerous but which it failed to warn the contractor about.
2. *Edward Lindenberg* v *Joe Canning Jerome Contracting Ltd* (1992) – contractor should have warned building surveyor about obvious dangers in plans.
3. *Brunswick Construction Ltd* v *Nowlan* (1974) (Canada) – contractor should have warned owner of defects in plans if they were so obvious that it should have recognised them.
4. *Victoria University of Manchester* v *Hugh Wilson & Lewis Wormersley and Pochin (Contractors) Ltd* (1984) 2 ConLR 43.
5. JCT63 Standard Form of Contract.
6. *Equitable Debenture Assets Corporation Ltd* v *William Moss & Others* (1984) 2 ConLR 1.
7. *Pride Valley Foods Limited* v *Hall and Partners and Hall Another (Contract Management) Limited* (Court of Appeal, 28 June 2001).
8. *Aurum Investments Ltd* v *Avonforce Ltd (in liquidation)* (2001) CILL 1729.
9. Recommendation of the editors of Building Law Reports at (2000) BLR 139.
10. *University of Glasgow* v *William Whitfield and John Laing (Construction) Ltd* (1988) 42 BLR 66.
11. *Bunge Corporation* v *GATX Corporation* – US – Louisiana Supreme Court.
12. *Kemper* v *Don Coleman, Jr., Builder, Inc.* No. 35, 172 CA (1987).

Estoppel

Probably the simplest way of explaining estoppel is by referring to the fact that a person who has made a promise must be estopped (stopped) from going back on it. Estoppel prevents a person from denying the truth of a statement it has previously made or of facts that it has led the other party to believe. An estoppel situation arises when one party, by words or conduct, leads another person to believe in the existence of a certain fact or situation and that other party has acted in reliance on that belief to its detriment. In these circumstances the party upon whose words or conduct the other party relied may be prevented (estopped) from denying the existence of the relevant fact upon which the other party relied.

Equitable estoppel can be divided for ease of understanding and use into ***promissory estoppel*** and ***proprietary estoppel***. In promissory estoppel the representor is precluded from going back on its representation or promise (this type of estoppel is most closely associated with contracts), whereas in the proprietary estoppel the representor is precluded from denying the other party's supposed rights in the representor's property. In effect proprietary estoppel is based upon misleading through encouragement and an apparent willingness to accept, whereas promissory estoppel is based upon misleading through representation.

The courts of equity developed proprietary estoppel to resolve unconscionable situations occurring with respect to land which can arise in two ways:

- where the land is improved by a party that believes it owns the land and the actual landowner is aware of this belief;
- where the land is improved by a party that has been led to believe by the landowner that it will obtain an interest in the land.

The usual example of proprietary estoppel is where a plaintiff expends money on land in the expectation that it shall have rights over it. In such a case equity will not permit the defendant landowner to enforce its legal rights and thus deprive the plaintiff of its expectation, where it would be unconscionable for the defendant to insist on those legal rights.

In accordance with *Willmott v Barber*,[1] the five elements put forward as being necessary to justify proprietary estoppel are:

- the party claiming proprietary estoppel has made a mistake as to its legal rights;
- that person has expended some money or done some act on the faith of that mistaken belief;
- the possessor of the legal right must know of the existence of its legal right which is inconsistent with that of the other party claiming proprietary estoppel;
- the possessor of the legal right must know of that other person's mistaken belief;
- the possessor of the legal right must have encouraged that other person in their expenditure of money or in doing other acts on which the other person relies, either directly or by abstaining from asserting their legal right (by acquiescence).

Promissory estoppel arises when a person makes a promise not to insist on, or enforce, its legal rights against another, or at least its representations to that effect were sufficiently clear and unequivocal to lead another to believe that those rights would not be insisted upon or enforced, and on the strength of such promise or representation the other party clearly and reasonably alters its legal position to take a course of action based upon that promise or representation. The promise or representation need not be supported by consideration in order to legally bind the person making it.

There are several other names used to describe the different categories of estoppel, the ones that the reader will most likely come across in an international context being:

- estoppel by agreement or convention;
- estoppel by representation;
- legal estoppel;
- estoppel by deed;
- estoppel by record.

Estoppel by agreement or convention applies where two parties have agreed on a base fact, which may be other than what the fact actually is, or where the convention for dealing between the parties clearly indicates that one party will allow a reasonable period of time to elapse before relying on its rights. Each party is estopped from coming back at a later date denying the facts as agreed. An example of the first case is if the parties agree on a certain exchange rate to apply to the dealings between them and they both act on that assumption, one party cannot return at a later date saying that the exchange rate should be something else.

The second case could arise for instance where the parties have an ongoing relationship and although payment should be made at, say, monthly intervals it has over a period of time been regularly made at two-monthly intervals. The person to whom payment is to be made will be estopped by convention from insisting on its rights to monthly payment.

Estoppel by representation relates to a representation made by one party to another that the latter relies upon. The person making the representation is estopped from later denying the truth of its representation.

Legal estoppel is estoppel as a matter of fact, which arises when there is a material fact that induces an action.

Estoppel by deed may arise when a seller does not have title at the time it purports to convey a property. If it subsequently acquires title to that property, it would be estopped from then denying that title and transferring it as originally represented. In the US, estoppel by deed is sometimes referred to as ***estoppel in writing***.

It is not unusual to hear equitable estoppel referred to as ***estoppel in pais*** or for ***estoppel by conduct*** to be used in place of both estoppel by agreement or convention and estoppel by representation. A further classification is ***estoppel by record*** that is not relevant in a construction context and covers matters arising from records of the courts in matters such as crime and probate, basically in that once a court has decided an issue then it cannot be re-litigated between the parties.

Estoppel is a 'rule of evidence' and cannot create a cause of action – it provides 'a shield and not a sword'. Countries that have adopted Civil law systems, including the French Civil Code, do not have anything to correspond to the 'rules of evidence' of Common law systems and so have no doctrine of estoppel.

1. *Willmott v Barber* [1880] 15 ChD 96.

Exclusive Remedies

It is common for standard forms of construction contract to include provisions to the effect that the various remedies provided in the contract, particularly in respect of extensions of time, additional costs and damages, are to be applied without prejudice to whatever Common law rights and remedies the parties may possess. It is equally common for non-standard contracts to include specific clauses to the effect that the contractual remedies represent the exclusive rights of the parties to the contract.

Where a contract provides express remedies for the normal contractual claims to be expected under a construction contract then an exclusive remedy provision may bar other Common law rights and remedies of the parties. It has been recognised by the courts in England and Wales that in general where the parties have agreed on a comprehensive set of contractual provisions covering remedies in relation to the whole spectrum of contractual claims they consider likely to occur, it is reasonable for them to exclude all claims which have not been expressly provided for in the contract.

The decision as to whether or not to bar all other claims is a commercial one to be taken by the parties prior to signing the contract as noted in the case of *Strachan and Henshaw* v *Stein Industrie*.[1] This case concerned clause 44.4 of MF/1, the model form of conditions for the supply of electrical or mechanical plant published by the IEE[2] and I Mech E[3] in the UK. This clause states:

> *the purchaser and the contractor intend that their respective rights, obligations and liabilities as provided for in the conditions, shall be exhaustive of the rights, obligations and liabilities of each of them to the other arising out of, under or in connection with the contract or the works, whether such rights, obligations and liabilities arise in respect of or in consequence of a breach of contract or of statutory duty or a tortuous or negligent act or omission which gives rise to a remedy at common law.*

Strachan had made arrangements for tea break cabin facilities to be provided for its workforce immediately adjacent to the site of a power station project in order that it could meet its statutory obligations as regards rest periods and so on. The employer instructed Strachan to move the facilities half a mile away from the site. Strachan argued that this instruction constituted a variation under clause 27. The arbitrator found for the employer and Strachan appealed.

The Court of Appeal held that the instruction was not strictly a variation 'of the Works' under clause 27 since 'Works' were defined elsewhere in the contract as referring to all plant to be provided and work done by the contractor under the contract. The court held that this definition was specific in its meaning and should not be distorted to include arrangements made by the contractor to facilitate its workers.

In upholding the arbitrator's decision the court recognised the commercial sense in providing expressly for the claims the parties intended to be allowed, and in excluding all possible claims other than those for which specific provision was made in the contract. In support of this view the court acknowledged that the form of contract MF/1 contained no less than 33 contractual provisions that covered all the normal contractual claims. They agreed that if the parties wanted to bar all other claims there was no reason why they should not do so, the contractual provisions being extensive in respect of those claims that were allowed for.

Where the strict application of an exclusion clause would give a party unfair or unreasonable benefits from its own acts of prevention, however, it would be unlikely that a court would agree to its implementation. Similarly a contract which purported to exclude all contractor remedies, including remedies otherwise available to the contractor upon an employer's acts of prevention or hindrance, would be construed contrary to the employer's interpretation.

Where a remedy provided for in the contract is dedicated, either individually or together with all other remedies provided in the contract, as an exclusive remedy then it is the sole remedy for the situation to which it relates. Conversely, a contractual remedy is not an exclusive remedy unless the contract expressly so provides. If a party promises a specific performance as an exclusive remedy, its failure to perform in the manner specified both vitiates the nature of the remedy and gives the injured party a right to demand specific performance, to the extent that is appropriate under the contract.

Exclusion clauses are common also in insurances, in particular in respect of workers' compensation claims. This aspect of exclusivity provides a considerable amount of case history in the US, largely to the effect that such clauses are upheld as barring all other claims. Various states in the US have applied their own modifications to this general principle and the applicable laws should be reviewed before proceeding with any claims in these jurisdictions.[4]

1. *Strachan and Henshaw* v *Stein Industrie* (8 December 1997), Court of Appeal.
2. Institute of Electrical Engineers, UK.
3. Institution of Mechanical Engineers, UK.
4. New York law provides three principal exceptions to the exclusivity doctrine in respect of workers' compensation claims concerning:
   - where an employer fails to provide the necessary insurance coverage;
   - where the injury arises out of an intentional act of either the employer or the employee; and
   - where a third party sued by an employee can only implead against the employer for a contribution or indemnification where the injury suffered by the employee was 'grave' (death, permanent disability etc.).

Exemption Clauses

There are basically three types of exemption clause in general use in the construction industry which purport to exclude or limit the liability of one of the parties for either breach of contract or some other potential liability (such as liability in tort), namely:

- those which exclude a party's liability altogether;
- those which limit the damages that can be recovered from the defaulting party;
- those which make liability subject to certain pre-conditions (for example, that a certain action must be taken within 12 months of the breach to which it relates).

For an exemption clause to be enforceable, certain conditions must be met:

- it must be validly incorporated in the contract; and
- its meaning must be clear and must match the nature of the defence to which it is to be put; and
- it must not be prevented by statute or by Civil Code of the country; and
- the contract must remain sufficiently intact that the clause still has some legal force.

If the exemption clause is included in a signed contract then it cannot be avoided unless there is evidence of fraud or misrepresentation, or unless it is rendered void or unenforceable by statute or by the Civil Code of the country.[1] The onus is on the enforcing party to prove that the exemption clause is part of the contract and on the defendant to prove fraud or misrepresentation or that it is void or unenforceable under statute or Civil Code.

The enforcing party will usually try to prove that:

- the other party signed a document containing the exemption clause (i.e. it is an explicit part of the contract); or
- the enforcing party took reasonable steps to inform the other party before the contract was signed that the exemption clause existed and applied to their contractual relationship (for example, a notice in a hotel bedroom excepting liability for the theft of valuables not kept in the hotel safe[2] would be an adequate warning of fact); or
- there was a consistent pattern of previous dealings between the parties which were both regular and consistent and which had included the exemption clause, such that incorporation of the exemption clause in the disputed context was clearly implied[3] (the tests to be applied to show 'regular and consistent' dealings would inevitably be more stringent when applying to businesses rather than to ordinary consumers).

The tests for reasonableness to be applied to an exemption clause in England and Wales are defined under the Unfair Contract Terms Act of England and Wales.[4] These tests would determine:

- the strength of the bargaining positions of the parties relative to each other, taking into account (among other things) alternative means by which the customer's requirements could have been met;
- whether the customer received an inducement to agree to the term, or in accepting it had an opportunity of entering into a similar contract with other persons, but without having to accept a similar term;
- whether the customer knew or ought reasonably to have known of the existence and extent of the term (having regard, among other things, to any custom of the trade and any previous course of dealing between the parties);
- where the term excludes or restricts any relevant liability if some condition is not complied with, whether it was reasonable at the time of the contract to expect that compliance with that condition would be practicable;
- whether the goods were manufactured, processed or adapted to the special order of the customer.

Assuming that the enforcing party successfully proves the legitimacy of the exemption clause, the other party will have little chance of defending its position. The one slim chance that it might have is to plead 'non est factum',[5] part of the doctrine of mistake, in the course of which it will try to prove:

- that it was unable, through no fault of its own, to understand the document without explanation (due to illness, lack of education or lack of understanding of the language in which it was written); or
- despite the fact that it took all reasonable precautions, the document signed was significantly different from the one it thought it was signing.

Obviously this plea would not be very convincing in respect of a contract for a major international construction project where the parties were all experienced in international construction and were each advised by their own teams of specialists in every relevant discipline, including lawyers.

If there is any question of ambiguity then the courts would apply the 'contra proferentem' rule[6] and would interpret the clause against the party who drafted the contract. The courts would also resist interpreting the clause in such a way that the defendant would avoid liability for a breach that was so fundamental as to deprive the other party of a substantial part of the benefits it was anticipating receiving under the contract.

The courts would look more favourably on a clause that limited liability rather than excluded it totally. In *Ailsa Craig* v *Malvern Fishing*[7] the owner of a vessel entered into a contract with a security firm, the contract containing an exclusion clause and a limitation of liability clause, each in respect of different circumstances. The vessel's owners suffered damage estimated at £55,000 as a result of the admitted negligence and breach of contract by the security firm. Even though the security firm admitted breach of contract, the limitation clause was held to apply in this instance as it reflected the true intentions of the contracting parties. The parties' intentions were considered not to extend to damage on this scale.

The interpretation of an exemption clause will take account of its natural and ordinary meaning read in the light of the contract as a whole, giving weight to the context in which the exemption clause appears and the nature of the contract. Negligence is a particularly complicated issue in relation to exemption clauses and such clauses require extremely clear drafting if they are to be upheld by the courts.

As noted above, an exemption clause excluding liability for negligence must be especially clear in its wording. It must state an express exclusion of liability for negligence. Where there is no express reference in the exemption clause to the exclusion of negligence the courts will have to decide whether the words used were in fact wide enough without any doubt or ambiguity to include such a reference. Negligence was not specifically mentioned in the exemption clause in a courier firm's agreement with its customer but it was held[8] that, since they were in the business of carriers, negligence was included by implication, the very nature of their business entailing an obligation to take care of the goods being transported – there were no grounds other than negligence which the exemption clause could reasonably have been intended to refer to.

An exemption clause would be denied by the courts where the service actually provided was radically different from that anticipated under the terms of the contract (i.e. where there was a fundamental breach of contract) resulting in damage for which the enforcing party was responsible. Taking another transportation case as an example, an exemption clause was included in respect of liability for any deterioration of goods being transported. The goods, quick-frozen scampi, were contracted to be shipped in a refrigerated hold of a ship, but in the event they were handled at atmospheric temperatures while waiting to be loaded and went bad. The exemption clause did not allow the defendant to escape liability.[9]

Exemption clauses may in some cases protect a party who is in fundamental breach of the contract but this is a matter of construction of the relevant contract clauses. England and Wales and Hong Kong are among countries which have legislation in place and which regulate the use of exemption clauses.[10]

In an interesting South African insurance case,[11] the insurer relied on the fraud exclusion clause in the general conditions of the policy it issued to a motorist that said:

> *If any claim under this policy be in any respect fraudulent, or if any fraudulent means or devices be used by the insured or anyone acting on his behalf or with his knowledge or consent to obtain any benefit under this policy, or if any event be occasioned by the wilful act or with the connivance of the insured, the benefit afforded under this policy in respect of such claim shall be forfeited.*

The insurer alleged that certain information contained in the claim form relating to the insured's version of the accident was false and that the general condition exclusion clause therefore justified rejection of the claim.

The court found that, contrary to the insured's version of the accident:

- it was not raining heavily at the time of the collision;
- a taxi was not involved in the collision;
- the taxi was not the cause of the collision;
- the whole description as to how the vehicles were damaged was untrue;
- the alleged speed of the appellant's vehicle was not true.

The information provided by the insured was false and was 'not according to fact, wrong and incorrect' and was not merely a version that differed from the version of the insurance witnesses. The statement made by the insured in the claim form was also found to have been made with full knowledge that it was false, with fraudulent intention designed to show that it was not negligent (although it was not necessary for the insured to have done that, since it was insured against its own negligence).

The court found, however, that the claim for the damage to the vehicle was not affected in any way by the false statement in the claim form, and the false statement did not make the claim itself a fraudulent one. The fact that the statement was knowingly made could not in the court's opinion reasonably have influenced the insurer as a prudent insurer in its review of the claim. It did not exaggerate the claim, the insured did not seek to recover something under the policy to which it was not entitled, and in that regard the statement, although false, was not material.

The court differentiated between the clause in the policy which referred to a claim being fraudulent and between the fact that the claim form itself was fraudulent. Since the insurer was the author of the policy, the contra proferentem rule was relied upon to resolve the ambiguity in the exclusion clause.

The court, however, denied the insured interest on the claim and also the costs of the action to show its contempt of the insured's behaviour.

Most national legal systems, whether in Common law or Civil Code jurisdictions, adopt the doctrine of freedom of contract and give regard to what each party could legitimately have expected from the performance of the contract. The French courts strike down unfair contract terms but work under a very broad definition of an unfair term, which results in them frequently having to consult with the Unfair Terms Commission (CCA) (an independent authority) in order to refine their definition on a case by case basis. The French Civil Code embodies and expands on the Common law principle that the 'cause'[12] cannot be false or unlawful (i.e. prohibited by legislation, contrary to public morals or to public policy[13]).

The following excerpt from the Dutch Civil Code teaches a salutary lesson to those who think that exemption (or other express) clauses in the contract will remain inviolate at all times regardless of jurisdiction:

> *...each party must always during its performance take into account the justified interests of the other party. This is an obligation of good faith which may entail that a party may not rely on an express provision in the contract, if under the circumstances that would be unacceptable under the requirements of reasonableness and equity...*[14]

Under the Dutch Civil Code (Burgerlijk Wetboek) 'reasonableness and equity' or 'redelijkheid and billijkheid' is paramount and will be ignored at your peril. The same principles will apply in jurisdictions whose Civil Codes are modelled upon the Dutch one, such as Indonesia, so be warned.

1. In England and Wales this could happen, for instance, where it fell foul of the Sale of Goods Act or the Unfair Contract Terms Acts. In Germany Article 444 of the Civil Code Book 2 Law of Obligations states 'The seller may not rely on an agreement excluding or restricting the buyer's rights in respect of defects if the seller fraudulently concealed the defects or if he has guaranteed the quality of the thing.'
2. *Olley* v *Marlborough Court Ltd* (1949) 1 KB 532.
3. *Henry Kendall & Sons* v *William Lillico & Sons* (1969) – previous dealings contained the disputed exemption clause.
4. Unfair Contract Terms Act 1977, Schedule 2 – 'Guidelines for the Application of Reasonableness Test'.
5. Literally 'it is not his deed' – a special defence in contract law to allow a person to avoid having to respect a contract that he or she signed because of certain reasons such as mistake.
6. See 'Contra Proferentem' section on p.64.
7. *Ailsa Craig Fishing Co. Ltd* v *Malvern Fishing Co.* (1983) 1 WLR 964.
8. *Thomas National Transport (Melbourne) Pty Ltd* v *May & Baker (Australia) Pty Ltd* (1966) 115 CLR 353.
9. *Clayton Love* v *B & I.*
10. (i) UK – Unfair Contract Terms Act; Unfair Terms in Consumer Contracts Regulations 1999 (complies with the European Union's Directive on Unfair Terms in Consumer Contracts, adopted on 5 April 1993). (ii) Hong Kong – Control of Exemption Clauses Ordinance enacted 1990 (referred to as 'CECO' and based upon the UK's 'Unfair Contract Terms Act').
11. *Strydom* v *Certain Underwriting Members* 2000 (2) SA 482 (W).
12. French Civil Code Article 1131: 'An obligation without cause or with a false cause, or with an unlawful cause, may not have any effect'.
13. French Civil Code, Article 1133.
14. Dutch Civil Code, Article 6: 248.

Fitness for Purpose

To establish whether a material or service provided is fit for its purpose it will be necessary to look into the particular circumstances of the contract under which that provision has been made. The requirement of 'fitness for purpose' exceeds the requirement for exercising 'reasonable skill and care' and imposes upon the contractor a strict obligation to succeed. If success is not achieved then the contractor is liable for the consequences of its failure to the extent that it would not escape liability for fitness for purpose by demonstrating that it had exercised reasonable skill and care.

The 'fitness for purpose' requirement must, however, relate to a specific purpose and the client must be proven to have placed reliance upon the other party, designer, contractor or whoever, to fulfil that requirement. This would be the case should such a clause be inserted in an international EPC/turnkey construction contract whether under Common law or Civil Code jurisdictions and we must constantly bear in mind the implications of such a clause throughout the execution of the contract.

Where the contractor is supplying materials in the course of building works, there is an implied warranty that the materials used will be of 'good quality and reasonably fit for the purpose for which the contractor is using them, unless the circumstances of the contract are such as to exclude any such warranty'.[1] There is also an implied warranty that the work as completed will be reasonably fit for the particular purpose of its intended use by the employer. The questions for us to ask in order to consider the implication of a duty of fitness for purpose are:

- In respect of materials supplied, was the contractor supplying materials in the course of its business?
- In respect of work executed, was the work of a kind that the contractor held himself out as performing?
- Did the employer make known the purpose of the goods and of the finished work to the contractor, either expressly or by implication?
- Did the circumstances show that the employer trusted the judgement of the contractor in these respects and not its own?

The court will look at the contract as a whole when interpreting the true meaning of the fitness for purpose provision but ultimately it is the result rather than the method of achieving it that will be decisive in determining whether fitness for purpose has been achieved. It has been held[2] that:

> *what they (the sub-contractors) had to do was to achieve a particular result, namely a support system which was fit for purpose.*

Case history supports the premise that, where the employer has specifically informed the contractor of the use to which the building it is contracting to construct is to be put, there is an implied term in the contract that the contractor will design the building accordingly.[3] In one case in particular the vibration of forklift trucks, which the contractor had been told would operate on the first floor of the building it was constructing, caused the floor to crack. The contractor was held to be in breach of an implied term that the design would be reasonably fit for the purpose intended. The claimant proved both purpose and reliance to the satisfaction of the Court of Appeal judges.

In an Australian case, a grain drying silo which the contractor designed and constructed was subject to water seepage into various below ground structures, rendering the silo unfit for the grain drying purpose for which it was intended.[4] The contractor was held liable under the 'fitness for purpose' principle since it had been told the exact purpose for which the silo was to be used and the employer had placed reliance upon its expertise to make it function correctly. The judge in the *Viking Grain* case qualified the fitness for purpose requirement by stating:

> *The obligation to design a product fit for its purpose is already tempered by the fact that only reasonable fitness is demanded.*

The bills of quantities or specifications may state with great precision that the contractor is to use particular materials of a certain type or from a certain manufacturer and the contractor must take heed of these requirements or else it will be in breach of the express terms of the contract and vulnerable for a breach of contract suit and a claim for damages by the employer. The contractor may, as a result of the implied term of fitness for purpose, find itself responsible for defects even when it is doing no more than what is required by the bills of quantities or specifications. It has been held that if the conditions of contract state something to the effect that 'nothing contained in the contract Bills shall override or modify the application or interpretation of that which is contained in the Articles of Agreement, the Conditions or the Appendix' then such a clause would prevent onerous clauses being added to the bills of quantities or bills of specifications imposing additional obligations on the contractor.[5]

Clauses in the bills of quantities or specifications in relation to materials or services can be too precise and employers should take care when drafting to ensure that they are not in fact reducing the contractor's obligations as to fitness for purpose.

An employer specified a particular hardcore for placing around foundations and provided for its testing and approval by the architect. The contractor used the hardcore as specified but it failed the tests. The courts held that the specific nature of the specification clause did not oust the implied terms that the fill was to be of merchantable quality but, since it was found to be of merchantable quality, and the employer had the means of knowing that it was not, however, of the standard required, the contractor was held not to be in breach.[6] The precisely-documented specification and procedures for testing and approval showed that there was no reliance on the contractor in selecting the hardcore and so no implied warranty for fitness for purpose and no liability for losses which flowed from the defective hardcore – a case of over specifying?

In summarising the application of this ruling we can say that if a defect arises from the contractor's compliance with the employer's specific design requirements the contractor is likely to be liable for merchantability quality but not fitness for purpose. Fitness for purpose liability could, however, arise where the contractor was responsible for the selection of materials unless the contract contains express provisions to the contrary.

The French Civil Code makes no specific mention of fitness for purpose but imposes an obligation that: 'Agreements are binding not only as to what is therein expressed but also as to all the consequences which equity, usage or statute give to the obligation according to its nature'.[7]

This leads to the same conclusions as those reached under Common law where the various agreements and sale and supply of goods and services statutes make or imply provision for a duty of fitness for purpose.

If there is a lesson to be learned, it is that, unless the terms (express or implied) in relation to work and materials are considered carefully, contractors may find themselves saddled with obligations far more onerous than they might otherwise have assumed. At the very least, the lesser responsibility of reasonable skill and care should be taken at all times.

1. *G. H. Myers & Co.* v *Brent Cross Service* [1934] 1 KB 46.
2. *John Lelliott (Contracts) Ltd* v *Byrne Bros (Formwork) Ltd* [1992] 31 ConLR 89.
3. *Greaves (Contractors) Ltd* v *Baynham Meikle & Partners* (1975) 3 All ER 99.
4. *Viking Grain* v *T. H. White Installations* (1985).
5. *John Mowlem & Co. Ltd* v *British Insulated Callenders Pension Trust Limited* (unreported).
6. *Rotherham Metropolitan Borough Council* v *Frank Haslam Milan and Co.* (1996) CLJ 333; 78 BLR 1.
7. French Civil Code, Article 1135.

Float and its Ownership

There has been much debate in international construction circles in recent years as to who exactly owns the programme float built into a CPN. 'Float' is best explained as the contingency time allowance between the date on which an activity is shown on the CPN as to be completed and the date when the contractor actually plans to complete that activity. There would be several individual activity floats and these could either be absorbed in the overall contract time allocation or could be taken through to the end of the contract period to give an overall contractor anticipated contract completion date somewhat earlier than the programmed contract completion date.

The question of 'who owns the float' is often raised when there are claims for extension of time or when the employer issues a variation order that somehow changes the time that the contractor requires to complete the works or part of the works. In the latter case the employer's argument will be that the additional work instructed under the variation order can be completed within the original contract period because of the available float. The contractor will argue that the float lies in its ownership and that it built it into the programme either as a cushion to protect it from delays for which it may be responsible during the course of the contract or as a time-saving mechanism benefiting it in a saving of overhead costs etc. for the reduced contract period, a saving which it would also have allowed for in its tender estimate.

It is generally accepted that the contractor will be entitled to an extension of time where the event which gives rise to a claim is a 'relevant event'[1] under the contract and will in fact delay completion beyond the contract completion date, as assessed at the time that the event occurs. The question of float must always be considered at the time when the delay occurred, before subsequent developments were allowed to cloud the issue. It is a basic rule of critical path scheduling that delays off the critical path will not give rise to an extension of the contract period until the total float along those paths has been used.

The logical approach to the question of 'who owns the float' is to view each case on its own merits. If the programme has a six-week end of contract float, is proceeding generally on schedule and then work is delayed by four weeks by a relevant event, the employer would not be obliged to grant an extension of time to the contractor. If the contractor, however, was at that time already in delay by five weeks due to a non-relevant event for which it was responsible, it would be logical for the employer to grant the contractor a three-week extension of time only, i.e. the difference between the delay caused by the relevant event and the then available float (original float less that already used up by the contractor). This is the 'first come first served' rule that suggests that the float belongs to the party that uses it first and should be applied unless there is an express provision in the contract that prohibits such an approach. This approach has actually been decided by the courts in the US as the correct approach to be taken.

Where, at the time that they enter into the contract, the employer is aware of the contractor's intention to complete before the contract completion date and where that intention was accepted as reasonable and realistically achievable, it has been suggested that, should a subsequent employer's delay affect the contractor's ability to complete by its planned completion date, the contractor would be entitled to the costs directly caused by the employer's delay even though the contract completion date was unaffected and no extension of time was granted.[2]

The approach in England and Wales is that the contractor cannot claim for damages for delay, only the direct costs incurred[3] whereas in the US it has been decided that, in a project delay situation rather than a particular path delay situation, the contractor would be entitled to recover damages for delay even though the revised completion date is still earlier than the contract completion date. To justify such a claim the contractor would be expected to prove that its originally intended completion date was reasonable and that, prior to the employer's delaying event, it was on track to achieve it.

Where a main contractor has a certain float in its programme and where it employs several sub-contractors to perform various works, it has been held that the main contractor must take available float into account when assessing the delay that a particular sub-contractor is responsible for. The main contractor cannot put the float aside and save it for another application while claiming recovery from the delayed sub-contractor of hypothetical losses it would have suffered had the float not existed.[4] The main contractor is not allowed such discretion.

Unfortunately neither the JCT or ICE standard forms of contracts in common usage in England and Wales deal with float in any way and the architect or engineer has total discretion under these forms of contract as to how to extend the time for completion. In general, where the programme is a contract document, the employer should grant an extension of time on the basis that it had accepted the programme and in so doing had accepted that the contractor could build into it a contingency or float to account for unforeseen difficulties. Where the programme is not a contract document, even if it is recognised under the contract as it is under the ICE contract, the programme is not contractually binding and therefore the employer would be justified in rejecting the contractor's claim to the float included within the programme. If the contractor cannot be bound contractually to the programme it cannot expect to maintain the float included in it.

It is possible for the parties to a contract to agree on the inclusion of a clause in the contract that defines clearly who owns the float and how it should be treated should the above or similar situations arise. Contractors have to make a positive decision when submitting their tender bids as to whether they should show float in their programme or keep it up their sleeve as a cushion for their own subsequent delays. If they adopt the latter approach they would have difficulty in claiming costs where the relevant event caused them to lose some portion of their own float without actually extending the contract period.

1. A 'relevant event' is an event that is stated in the contract as obliging the employer to grant the contractor an extension of time in relation to it.
2. Society of Construction Law 'Delay and Disruption Protocol'.
3. *Ascon Contracting Ltd* v *Alfred McAlpine Construction Isle of Man Ltd* (1999) 66 ConLR 119.
4. *ibid.*

Fundamental Elements of Effective Contract Management

The principal intention of this book is to arrive at a state where we, whether we are architects, surveyors, project managers, employers' representatives, engineers or claims consultants, whether we are acting for employers, contractors or sub-contractors, are capable of achieving the highest level of effective contract management possible. In this book in general, and in this section in particular, I use the pronoun 'we' to represent us all, both collectively and individually and the term 'contract management' to represent our main function, albeit from commercial and contractual standpoints rather than a technical one.

The fundamental elements that combine to bring about truly effective contract management can be defined under ten principal headings:

- contract awareness – thorough review and understanding of the contract;
- document control (including correspondence, minutes, reports, photographs and so on);
- compliance with all the commercial/legal requirements of the contract (notices etc.);
- relationship management – maintain good relationships between client, contractor, joint venture and consortium partners, sub-contractors, consultants, etc.;
- change order control;
- claim management;
- programme management;
- sub-contractor and supplier control;
- material control (shipping, material inspections, customs clearance, insurance etc.);
- quality control.

Each of these issues needs to be given our closest attention in order that we are not only aware of everything that is happening in respect of a particular project at any given time but also, probably more importantly, so that we are able to influence events in order to support our overall contract management strategies.

I have found that most Western-based contractors and employers understand, or are flexible in their approach to, the concept that, if we are to fulfil our role effectively, we must be afforded open access to all aspects of the work being undertaken on the contract. Unfortunately this is a concept totally alien to those operating from the Far East and South-East Asia. The Japanese in particular are suspicious of sharing information between their own nationals in the various departments in their own organisations, let alone with a foreigner, which leads to a heightened need for diplomacy, persuasion and imagination in the efforts we employ in ensuring that we have as much of the required information as is at all possible under the particular conditions in which we are working.

We will now look at these subjects in more detail.

Contract awareness

The review of the contract should first be carried out at tender stage. At that time there will be a thorough examination of all the clauses proposed together with the specifications, all appendices and attachments and any other documents forming part of the contract. A cross check will be carried out in order to ascertain if there are any ambiguities and/or discrepancies between the various documents and questions, and deviation lists, or whatever else is required by the Tender Instructions, will be prepared and submitted to the employer in order to resolve any possible conflicts prior to submitting the tender. Ideally the project team should visit the site to see for themselves if there are any conditions there that could have a commercial impact upon the successful completion of the contract.

After the contract is awarded, we must familiarise ourselves with the contract, with the advice given pre-tender and with the actions taken as a result of that advice, if we are not already familiar with this from our involvement at the bid stage. We will then prepare our strategy based upon our review of the contract documents, scheduling times when information has to be provided, actions of any sort have to be initiated, submissions made and so on. This will be our matrix for the management of daily project events.

Document control

Successful contract management requires close attention to be paid to all forms of documentation, in terms of preparation and submission as well as in respect of document retention and maintenance in case of later disputes arising.

Documentation in this context does not only refer to the contract documents and the correspondence between the parties to the contract. It would include such items as minutes of meetings, telephone logs, site diaries, memos, job-site logs, photographs, monthly reports, method statements, QA/QC information including test procedures and results, certificates, notices, shipping information including customs duties, taxes and bills of lading and so on., permits and licences, change orders, survey information, re-measurement data, sub-contractor and supplier quotations and purchase orders, delivery notes etc.[1]

We will need to put procedures in place to ensure that, where we are not initiating it, we obtain copies of all documentation promptly and completely, and that documents are registered and maintained in an easily accessible manner. Wherever possible we should have the opportunity of reviewing all documentation not initiated by us, including outgoing correspondence, reports, meeting minutes, memos and so on to ensure that they comply with the contract conditions and with the company's project strategy, which we will have agreed and revised from time to time with the appropriate senior management.

Compliance with contract requirements as to notices etc.

It is essential that all notices required by the contract be given within the notice periods stated in the contract. In some contracts, notably the new FIDIC Silver Book for use on EPC/Turnkey projects, notices must be served within a set number of days and must also be replied to within a set number of days. If the contractor fails to serve notice under the Silver Book contract within the specified time period it will lose its right to claim.

It is sensible to prepare a schedule identifying each clause that has a notice period attached or a period within which information has to be provided, replies given, action taken and so on. This should be copied to the project team, who should be appraised of the consequences should the proper procedures not be followed. This schedule will act as a useful tool in the monitoring of site events.

Relationship management

It is extremely important for us to establish good working relationships with our own project team, with the other party to the contract and their project team and with the consortium or joint venture partners, sub-contractors, suppliers, etc. Cooperation between all the parties is essential for the smooth running of the project; unfortunately this does not always happen, or if it does it tends to peak at the early 'honeymoon' stage and then wane off as the project carries on and as problems arise.

Apart from the day to day working relationships that will develop over the period of the contract, it is essential that there are comprehensive contractual agreements in place between the consortium or joint venture partners, assuming that this is the way the project is set up. We should have input into the preparation of any consortium, joint venture and finance agreements which the party we work for is involved in and should ensure that such agreements are well balanced as regards the responsibilities and obligations of the parties under the agreements and also in relation to their application to the main contract. These should be agreed in full prior to signing the main contract to avoid later disputes, the resolution of which could mar the good relationships that have hopefully been built up between the parties during the bid and negotiation stages.

Being friendly, professional, fair and ethical in one's relationship with members of the project team, whether they be the labourer on site, a joint venture/consortium partner or the president of the employer company, can only benefit the project and thus increase the chances of future work. Having a good relationship with all members of the project team will ease the anxiety when solutions have to be found, often as a matter of urgency, to the inevitable problems that will occur on site.

Change order control[2]

Most contracts have provisions for the employer to issue change orders when it wishes to vary the works in some way. Reviews at the contract negotiation stage should ensure that there are also procedures in place whereby the contractor can request a change order, either where it is suggesting a change that will save the employer time or money or where some other situation is discovered or event occurs that it considers necessitates the issuance of a change order. All parties must follow events on site in respect of design, engineering, purchasing, QA/QC, construction, sub-contracting and so on to identify potential change situations and to serve the requisite notices in a timely manner in accordance with the specific requirements of the contract.

The contractor's project team must be made to understand that they must not act on any verbal instruction from the employer or the employer's representative that constitutes a change until such verbal instruction has been confirmed in writing by the employer. Confirmation must be both requested and received before any changes are implemented. It is helpful to have a clause in the contract that provides for the contractor to confirm the employer's oral instructions within a specified period of time, say seven days, of the employer issuing them and for their deemed approval where the employer fails to withdraw them within a further specified time following the contractor's confirmation.

Ideally the employer and the contractor should be able to agree the cost effect of a change order before commencing work on it and if this is not possible there should at least be a mutual understanding on the basis for such agreement, be it at contract bills of quantities or schedule rates, on a day-work basis or on a lump sum basis. The contractor should ensure at tender stage that there is a clause in the contract precluding the employer from omitting more than 20 per cent of the contract works, either in an individual change order or collectively in several, without the contractor then having the opportunity to renegotiate the rates or prices for the balance of the works, thus preserving its overhead and profit return on the project.[3]

Claims management

Claims management is dealt with in more detail in other sections of this book. This is an extremely onerous aspect of our job and will occupy much of our time. It can also be a very rewarding aspect of our job, both in respect of successful claim avoidance and of claim settlement, when at the end of the project we can look back on our achievements with satisfaction as to the manner in which we managed the contract.

Claims management incorporates several aspects of contract management that have already been addressed in this book and more that will be addressed later. This section gives the opportunity for a brief explanation of what is meant by claims management, and it is stressed that successful claims management should not be judged by the number of claims that have been submitted or rejected but on the manner in which any contentious issues have been resolved. It is my strong belief that there is as much job satisfaction and pride in successfully practising claim avoidance as there is in reaching a settlement on a disputed claim.

I was once interviewed for a position with one of the UK's so-called Claims Consultancies and was asked what I considered to be the most important attributes of a successful contract/claims manager. The reply I gave stressed, amongst other attributes, the importance of diplomacy, at which point the senior partner interviewing me shook his head in despair and said that he could see no room for diplomacy in claims management. That was the end of the interview as far as we were both concerned and no job offer was forthcoming, nor would one have been accepted if it had been offered as I stand by my opinion.

Claim avoidance, either to the extent that effective contract management precludes the need for claims to be initiated or at least to the extent that negotiations on claims can reach an amicable solution before they reach the stage where arbitration or litigation cannot be avoided, maintains better relationships between the parties (which can lead to future work and which certainly results in a less stressful working environment, which in itself results in a greater chance of claim avoidance being possible) and reduces the costs which would otherwise inevitably be incurred at arbitration or litigation. This cannot be achieved without more than a little diplomacy. Surprisingly the claims consultancy concerned is still in business, although I consider that its now well-known approach to claims does nothing to enhance the stature of the profession within the industry.

Claims management starts from the very moment that the tender documents are received. As stated above, successful claims management entails claim avoidance as much as claim initiation and should not be viewed by either the employer or the contractor as a specifically confrontational issue. The issues involved in claims management can be stated in several different ways at different stages in the project's life:

At tender stage:

- preparation of, and clear response to, deviation lists and query sheets in an attempt at claim avoidance (i.e. the clarification of ambiguities, unclear issues, conflicting descriptions etc. at this stage will avoid later disputes or disagreements arising between the parties).

During the execution of the project:

- complying with all contract requirements in particular as regards submission of, and approval of, information by predetermined dates or within periods set down in the contract;

- serving appropriate notices within the notice periods required by the contract;
- ensuring that there are clear routes of communication between the parties so that issues can be raised and discussed at the earliest opportunity;
- maintaining records of all events which might lead to a claim for additional time and/or cost;
- recording in writing any oral instructions which could lead similarly to a time or cost claim;
- maintaining accurate minutes of meetings, preferably countersigned at the time or soon after by the principal attendees as a true record of the meeting;
- ensuring that there are accurate records of all schedule changes;
- ensuring that the monthly report to the employer contains, amongst other items listed elsewhere in this book,[4] an updated schedule, an account of why there have been changes to the previous schedule, a listing of all change orders and claims issued within the previous month with costs or estimates of costs wherever possible, and a summary of correspondence issued during the month of the report.

At contract closure stage:

- reviewing the contract cost against the contract budget, identifying cost over-runs and cross checking against programme over-runs, change orders, dayworks, back-charges etc. to establish the reasons and whether or not the additional costs can be claimed.

If we are appointed primarily in a claims management role, it sometimes happens that our involvement in a project does not commence until some way into the contract period, usually when a commercial problem or claim issue appears which needs our particular attention. We then have to delve back through all the appropriate records, correspondence etc. to trace the events that have given rise to the issue in question together with the resultant effects. At this point everybody involved in the exercise will appreciate our concerns regarding the need for complete and accurate records and documentation to be maintained from the earliest stages of the contract. It will probably be too late for total claim avoidance although there will hopefully still be some scope for negotiation, and our duties in this situation will be to:

- identify and develop claim issues from available records;
- carry out a factual and detailed schedule analysis;
- assess damages (time and cost);
- prepare a claim report for initial negotiation purposes;
- serve notice of arbitration or other appropriate dispute resolution procedure where negotiations fail;
- prepare formal claim documents, brief and support lawyers at arbitration or other appropriate dispute resolution procedure.

The whole issue of claims management is closely tied in to many of the individual topics discussed elsewhere in this book and hopefully the reader will have a better grasp of the complexities of claim management by the time the end is reached.

Programme/schedule management

Programme management plays a critical part in the effective management of claims and disputes. It is essential that regular revisions of the project programme are submitted to all parties as a record which will be invaluable should a claim or dispute situation arise.

The delay analysis that will be undertaken as part of the claims management exercise will rely heavily upon regular programme reviews having been undertaken during the execution of the project. As explained under the heading 'Delay Analysis',[5] there are several different formats of programme that will be required, including the original planned programme and the final as-built programme. In addition there might be a need for 'as-planned but for', 'as-built but for' or 'as-built impacted' programme comparisons to identify the effects of alleged delaying events.

The employer and the contractor should each ensure that the preparation of the appropriate programmes and programme revisions is carried out as an on-going exercise throughout the contract and is not left until the end, by which time memories will have faded and project staff will have moved on to the next project.

Sub-contractor and supplier control

The contractor must exercise control over the commercial aspects of its sub-contractors and suppliers employed on the project to ensure, amongst other things, that their terms and conditions are compatible with those that the contractor has with the employer.

The contractor should endeavour to introduce back-to-back conditions, passing on contractor's risk to the sub-contractor or supplier wherever it is appropriate and possible to do so. This is particularly important in respect of bonds and guarantees, payment procedures, delays, changes to the works, insurances, warranties, confidentiality, IP rights and so on.

The contractor should ensure that prospective sub-contractors prepare their bids in a proper manner and should encourage them to visit the site and to clarify any potentially contentious issues before their bid is accepted and certainly before their sub-contract agreements are signed. To this end it is advisable for the contractor to oblige prospective sub-contractors to sign a form confirming that they have sought information on, discussed and agreed all such items and that their tender makes due allowance for these issues,[6] even if the terms of the sub-contract or purchase order state or imply this to be the case.

When there are claim issues involving sub-contractors or suppliers the contractor must decide whether they are solely for it to resolve directly with the sub-contractor/supplier concerned or whether they can be passed on to, or incorporated in the contractor's claims against, the employer.

In the latter cases it is up to the contractor to ensure that final settlement arrangements will be on a back to back basis, i.e. the contractor will fight the sub-contractor/supplier's case against the employer but the sub-contractor/supplier will have to accept the employer's decision or the result of subsequent mediation, arbitration, litigation etc. as the final one without any further recourse against the contractor. If, in situations where we are working for the contractor, we are able to have input into the preparation of the sub-contract agreements and purchase orders at negotiation stage, such a condition can be pre-agreed, saving argument when any subsequent dispute situation arises.

Back charges[7] against sub-contractors and sub-vendors must be vigorously pursued and the contractor should monitor potential back charge situations closely, establishing a register of back charges that will enable them to be properly recorded, tracked and settled. Once a back charge has been identified the contractor should initiate measures to keep separate the costs of all work executed in connection with the back charge and must ensure that sufficient funds are available on the account of the sub-contractor or sub-vendor concerned so that the back charge can be set off against monies otherwise due.

Material control

Again, when we are working for the contractor, our principal involvement in material control will be in respect of shipping, delivery schedules, customs clearance, material inspections and insurance claims.

In this respect we will monitor shipments against the programme and identify any delays or additional costs which have been incurred and for which the contractor is entitled to seek recompense. We will need to receive material inspection reports when materials are inspected upon delivery and ensure that appropriate action is taken to remedy any shortages or damaged items.[8] The contractor might have grounds for claims against the suppliers (inadequate packing), the shippers (storage on deck instead of below deck causing damage from the elements) or against the insurers (loss or damage) and these claims must be initiated within the appropriate time period and be accompanied by the relevant paperwork (bills of lading, commercial invoices, inspection reports, damage surveys and so on).

The other aspect of material control which we will inevitably get involved in is in relation to import taxes, duties etc. The contract will state who is responsible for their payment and might give directions for reimbursement where these are initially paid by the contractor but are the ultimate responsibility of the employer. We will need to monitor the rate of tax or duty paid and claim back any increases in such rates that have occurred since the tender base date, if this is permitted under the terms of the contract.

Quality control

Quality control is a critical part of the construction process and each party to the contract will have its own designated team of specialists dealing with this. The QA/QC team will be involved in the preparation, or approval, of inspection and test procedures, method statements and quality audits. We will need to monitor these events closely to ensure that the contractor's submissions are made in a timely manner in accordance with the requirements of the contract and that they cover all the aspects of tests, inspections and controls required by the contract. Similarly, we must monitor the employer's approval procedures to ensure that the employer responds promptly to the contractor's submissions and that it does not insist on more onerous procedures and inspection/test plans than were anticipated by the contract, thereby exposing itself to risk of the contractor submitting claim notices to recover the additional cost and/or time involved.

There is always an issue regarding inspection of materials off site, possibly in countries other than that in which the project is being constructed. The contract documents should be clear about who is responsible for the cost of tests and for the expenses of the visiting inspectors. We must monitor this situation closely to ensure that additional costs are not incurred, such as by the employer requesting additional tests not included in the contract or by the contractor delaying previously arranged tests, disrupting the arrangements of the employer's team, and to ensure that, if additional costs are incurred, they are reimbursed.

The employer should be particularly careful to ensure that its inspectors carry out their duties in a timely manner, thus preventing the opportunity arising for the contractor to submit a claim in respect of additional time and costs incurred as a result of their tardiness where, for instance, this has contributed to delays in completion of manufacture or has delayed shipment to site.

In all probability there will be a condition of final acceptance that requires a full dossier of QA/QC documentation to be lodged with the client prior to issuance of the Final Acceptance Certificate. The contractor must monitor the situation and ensure that its QA/QC department honours this requirement and does not delay either the issuance of the Final Acceptance Certificate or of any interim monthly or milestone payment certificate, where the submission by the contractor of QA/QC documentation is a prerequisite to its issuance.

If the procedures discussed above are implemented in full, then the efficiency with which the contract is administered will be recognised and appreciated by all the parties involved in the project, producing a feeling of comfort and goodwill that will permeate every aspect of the works. The various parties to the contract are each less likely to pursue frivolous claims against another party where the other party is known to have managed the project effectively, served and responded to notices in a timely manner, maintained accurate records throughout the contract, initiated only realistic and well-supported claims at the appropriate times during the course of the project, and responded to claims in a reasonable manner.

1. See complete list of documentation in the 'Document Control' section on p.102.
2. See 'Change Order Management' section on p.33.
3. See 'Omission of Work' section on p.189.
4. See 'Monthly Reports' section on p.177.
5. See 'Delay Analysis' section on p.86.
6. See Appendix 4, p.289 – 'Contractor's Pre-Tender Information Confirmation' form.
7. Back charges are, in this context, additional costs incurred by the contractor arising from:
   - default by the sub-contractor or sub-vendor under its sub-contract with the contractor;
   - actions of the sub-contractor or sub-vendor which result in damage, loss or additional expense being suffered by the contractor, employer or other sub-contractor or sub-vendor; or
   - services provided by the contractor outside the scope of work contained within the sub-contract (such as supply of electricity, craneage, labour, materials etc.).
8. Typical 'Overage, Shortage and Damage (OSD) Report' is shown in Appendix 6 on p.293. This can be adapted to suit each particular project.

Implied Terms

An implied term, unlike an ambiguous, conflicting or inconsistent term, is a term that is not expressed by the parties (either in writing or orally) but which the court 'reads into' the contract.

It is common in construction projects for one party to assume from its interpretation of the contract that the other party has a responsibility or obligation that the other party denies having. At this point the party claiming the responsibility/obligation against the other asserts that, although not expressly stated within the contract, such a responsibility/obligation is implied. Then the argument begins.

In *BP Refinery v Shire of Hastings*,[1] Lord Simon stated that:

> *For a term to be implied, the following conditions (which may overlap) must be satisfied:*
>
> 1. *it must be reasonable and equitable;*
> 2. *it must be necessary to give business efficacy to the contract, so that no term will be implied if the contract is effective without it;*
> 3. *it must be so obvious that 'it goes without saying';*
> 4. *it must be capable of clear expression;*
> 5. *it must not contradict any express term of the contract.*

The word 'implied' can be defined as meaning 'known indirectly' and can be something that is 'made known by analysing the surrounding circumstances of the actions of the parties involved'.

An 'implied term' can be either a condition or warranty[2] that must be implied in order to give effect to the presumed intentions of the parties. In some cases, the court implies a term on the ground that the parties probably had the term in mind when agreeing the contract terms but simply neglected to express it. Thus the court is giving effect to the actual intentions of the parties.

In some cases the court implies a term that the parties probably did not have in mind but would have wanted to include if they had thought of it. The justification for implying the term is that the court is simply giving effect to what the parties would have wanted had they bothered to address the topic in their negotiations. The court will imply a term if the language of the contract and the circumstances under which it was entered into give rise to the inference that the parties must have intended the term in question to be a part of their contract.

Terms can on occasion be implied by Common law, statute or Code. For instance, it is an implied term under every contract to which the Late Payment of Commercial Debts (Interest) Act 1998 applies in England and Wales that any 'qualifying debt' under the contract carries statutory interest. A qualifying debt arises where:

- there is an obligation under the contract to pay the whole or any part of the contract price; and

- that amount has not been paid on the agreed payment date or, if no date has been agreed, 30 days after the invoice or delivery of goods or services, whichever is the later.

Interest under such an implied term is treated in the same way as interest arising under an express contract term and, if not paid, can be sued for. The Act can only be contracted out of if there is a substantial contractual remedy for late payment of debt. A remedy shall be regarded as a substantial remedy unless:

- the remedy is insufficient either for the purpose of compensating the supplier for late payment or for deterring late payment; and
- it would not be fair or reasonable to allow the remedy to be relied on to oust or (as the case may be) to vary the right to statutory interest that would otherwise apply in relation to the debt.

Similarly in Hong Kong, the Sale of Goods Ordinance implies certain terms into each contract to which that statute applies, again whether the actual statute is specifically stated as applying to that contract or not.

Terms may be implied by obligation. An implied term arises because the nature of an agreement imposes certain legal obligations, and they must be allocated to one party or the other. In a residential dispute,[3] landlords were held to have an implied obligation to maintain common areas of a block of flats that was let in separate units, even though the tenancy agreements made no mention of this. It was held that someone had to shoulder this burden and, as the landlords retained control of these areas, it was held to be their responsibility.

Terms may be implied by custom, such as an implied term in a contract that arises from the custom in a particular trade or a particular locality. For example, it was held in a case in England[4] where there was a question regarding the sale of 1,000 rabbits that it was the custom in the area that 1,000 rabbits meant 1,200 rabbits. The rationale is that the parties did not express all of the terms of their agreement but rather only those that were necessary in that specific case, relying upon the regular customs of their trade for the remainder of their terms. The same principles would apply in a construction or engineering context; however, the party that intends arguing in favour of such an implied term must show that the custom is well established in the relevant industry, country or locality.

The term implied by custom must be consistent with any expressed term, and so it is within the parties' power to contradict the custom (in so far as it is legally permitted to in the particular jurisdiction concerned) by including an express term to that effect in the contract.

Terms may be implied by fact, such as where an implied term becomes essential to a contract because the contract simply would not work without it. The leading case in this respect, *The Moorcock*,[5] formulated the 'business efficacy' test, as mentioned in item (2) of Lord Simon's statement in the case of *BP Refinery (Westenport) Property Limited* v *Shire of Hastings* noted earlier in this section. This case concerned wharfingers who contracted for the claimants' ship to use their jetty. The ship was damaged at low tide and the defendants were held liable, on the basis that the parties must have presumed that it would be safe for the ship to berth at low tide. The acknowledgement of this implied term was held as being necessary in order to give business efficacy to the contract and to repair an obvious oversight.

A court will not be sympathetic to a party who wants to imply a term simply because it would make the contract 'more fair' or because the alleged implied term would be 'reasonable' under the circumstances. Rather, the court must be satisfied that the term is truly necessary for business efficacy or 'so obvious' that nobody would doubt its existence as part of the contract.

Situations in which a party is arguing that a 'significantly major' term must be implied into the contract must be treated with care because if a significantly major term is missing then there is a good chance that the court will hold that no contract was formed. This is especially true if it is not 'obvious' what term should be implied to fill the void left by the missing term (for example, if there are equally strong arguments supporting different ways of filling the void).

The employment relationship is another field where courts have been known to imply terms. It has been held by the courts, for example, that there are implied terms in employment contracts imposing upon the employees a duty of loyalty to their employer and a duty requiring them to act in the employer's best interests. It has also been held that there is an implied term imposing upon the employer a responsibility to take reasonable care not to injure its employees' health.

Normally, a term would not be implied if it would clearly contradict an express term. However, the powers of parties to 'exclude' an implied term through an express clause in the contract are limited and some clauses that profess to exclude liability are either not enforceable under any circumstances or are only enforceable under certain circumstances.

The above principally describes the situation in Common law jurisdictions, in particular those of England and Wales and Hong Kong, although most Common law jurisdictions apply the same doctrines, adapted as may be necessary to accord with their own particular legal enactments. The US 'Uniform Commercial Code' is a good example of the US attitude to implied terms following that of England. The French Civil Code system on the other hand is designed to rely upon the terms of the contract rather than the implied terms. An English court faced with a contract dispute will start its deliberations with the question 'To what did the parties agree?' whereas a French court under similar circumstances will ask the question 'Into what type of contract did the parties enter?' A Common law lawyer will attempt to resolve the dispute with an implied term, a Civil law lawyer with a rule.

The only concession under the French Civil Code to anything approaching an implied term is in the provision that says:

> *Agreements are binding not only as to what is therein expressed, but also as to all the consequences which equity, usage or statute give to the obligations according to its nature.*[6]

This limits the implied term to one arising out of equity, usage or statute and is therefore not as broad as the English interpretation, which goes as far as to include what the parties intended when they entered into the agreement.

1. *BP Refinery (Westernport) Property Limited* v *Shire of Hastings* (1978) 52 ALJR 20.
2. A *warranty* is a lesser term. Breach of a warranty gives the other party the right to sue for damages but *does not* give it the right to terminate its own performance. A *condition* is a very important term. Breach of a condition gives the other party the right to treat the contract as terminated, cease its own performance, and sue. An *intermediate (or innominate) term* is a more recently recognised class of term that can be either a warranty or a condition. The legal consequences of a breach of an intermediate or innominate term depend upon the impact upon the innocent party and do not follow automatically from prior classification as 'warranty' or 'condition'.
3. *Liverpool CC* v *Irwin* (1976) 2 All ER 39.
4. *Smith* v *Wilson* (1832).
5. *The Moorcock* (1886–90) All ER Rep 530; 14 PD 64.
6. French Civil Code, Article 1135.

Indemnity

Indemnity is an expressed or implied contract, promise or obligation given by one party to another stating that the obligor will compensate or reimburse the obligee for possible losses or damages of a particular type. It is in effect a form of insurance from one party to another party, and is another means of identifying and apportioning risk. In *Total Transport* v *Arcadia*,[1] Staughton LJ stated that:

> *The word 'indemnity' is ... used in two senses. It may mean simply damages awarded for breach of contract ... alternatively the word 'indemnity' may refer to [compensation for] all loss suffered which is attributable to a specified cause, whether or not it was in the reasonable contemplation of the parties.*

Each party should be covered by a clause in the contract whereby each party indemnifies the other against particular events stated in that clause. It is usual for the contractor to indemnify the employer against any suits, actions or administrative proceedings, claims, demands, losses, damages, costs and expenses etc. in respect of the death or injury of any person or loss or damage to any property (other than the works) arising out of or resulting from the performance of the works and caused solely by the negligence of the contractor, its employees, agents or sub-contractors. The employer in turn usually indemnifies the contractor similarly in respect of its negligence and that of its employees or agents.

Other indemnity provisions which might well be incorporated in an international construction contract and would usually be expressed in specific terms in addition to the general indemnity provisions could include such issues as Professional Indemnity and Medical Indemnity, each of which would have to be covered by a specific form of insurance.

Indemnity, in its general application, can be covered by an Indemnity Bond indemnifying an obligee against loss or damage that may arise as the result of a failure by the principal to perform as required under the contract. The indemnity could be met by reimbursement of a loss or financial damage, or repair/replacement of an item of property physically damaged by the negligence of the concerned party or one of the class of persons for whom it is responsible under the contract. We should ensure before commencing work on the project that the parties to the contract are protected against risk in this respect by means of an indemnity bond, insurance or other protection as might be required under company regulations. This would apply strictly as a contract provision irrespective of whether the contract was under Common or Civil law jurisdiction.

The key distinctions between a guarantee and an indemnity must be clearly understood:

Indemnity:

- Not conditional – the indemnifier is personally liable as primary debtor.
- Generally two parties to an indemnity – the creditor and the debtor.

Guarantee:

- Conditional – it is a subsidiary liability arising only if the primary debtor does not perform its obligations.
- Generally three parties to a guarantee – creditor, principal debtor and secondary debtor (i.e. the guarantor). The basic function of a guarantee is to provide an additional party who may be sued for performance, particularly where the primary debtor becomes insolvent.
- A guarantee is looked upon as being particularly advantageous where there is a high risk of insolvency of the primary debtor, in which case a guarantee gives more comfort than an indemnity.

This distinction was explained many years ago in *Birkmyr* v *Darnall*[2] along the following lines:

> *If A agrees to supply goods to B on the strength of C's promise that 'if he does not pay you, I will', C undertakes a secondary liability for the price, dependent on B's failure to pay. That is a contract of guarantee between A and C. But if C's promise is 'I will see to it you are paid', C is providing a separate undertaking as principal and therefore is providing an indemnity. In the first case, C's liability is dependent on B's default while in the second it is not. The essential point of the indemnity is that it is a distinct, autonomous undertaking. C's payment undertaking is a primary obligation which exists independently of the contract between A and B, which is governed by its own terms and which A can in principle enforce against C whether or not A has an enforceable claim against B.*

In another more recent case[3] it was explained that:

> *An indemnity is a contract by one party to keep the other harmless against loss, but a contract of guarantee is a contract to answer for the debt, default or miscarriage of another who is to be primarily liable to the promisee.*

A few other pertinent points to bear in mind in relation to indemnities, guarantees and insurances are:

- liability under an indemnity clause may be viewed as a debt rather than damages;[4]
- where the indemnity obligation is viewed as a debt there is no duty to mitigate loss;
- where viewed as a breach, the wording of the contract clause will determine whether or not there is a duty to mitigate loss;
- indemnities stand alone, taking out insurance cover merely provides added protection;
- the limitation period for an indemnity runs from the date the loss is suffered, not the date of breach;
- distinction between an indemnity and a guarantee will depend upon the construction of the words used in the text, and the fact that one or the other label

may have been used is inconclusive as regards the distinction;[5]
- guarantees need to be in writing.[6]

1. *Total Transport Corp.* v *Arcadia Petroleum Ltd* (1998) 1 Lloyd's Rep. 357.
2. *Birkmyr* v *Darnall* (1704) 1 Salk. 27.
3. *Yeoman Credit Ltd* v *Latter* (1961) 2 All ER 294.
4. *Jarvis* v *Harris* (1996) CA.
5. *Yeoman Credit Ltd* v *Latter* (1961) 2 All ER 294; *Motemtronic Ltd* v *Autocar Equipment Ltd* (unreported, 20 June 1996), Civil Division.
6. *Actionstrength* v *International Glass Engineering* (3 April 2003), HL.

Insolvency

There are numerous specialist books concerning insolvency written by experts in the field. This section will limit discussions to the basics of the subject sufficient to give the necessary background to help us understand the aspects of insolvency that we are most likely to encounter on an international construction project.

Insolvency is the condition a company is said to be in when it either does not have sufficient assets to cover its debts (i.e. liabilities greater than assets) or is unable to pay its debts as they become due. Insolvency is a prerequisite of bankruptcy.

Under the Insolvency Act of 1986 of England and Wales there are several options open to insolvent companies, namely:

- Administration – temporary delay of action (moratorium) by creditors to allow company to restructure – this needs to be initiated by a court order but this can usually be obtained quite quickly and has the benefit of immediately putting a stay on the actions of creditors.
- Compulsory Voluntary Arrangement (CVA) – a voluntary arrangement between a company (which need not be insolvent) and its creditors under the supervision of an insolvency practitioner. This is usually cheaper and simpler to arrange than an Administration, the down side being that there is no stay on the actions of creditors during the CVA process.
- Receivership – the appointment of a receiver, usually an insolvency practitioner, by a creditor, or by the holder (such as a bank) of a floating charge over the company's assets, (in this case called an administrative receiver[1]) to collect the assets in and realise those assets and repay the debts.
- Liquidation –winding up a company voluntarily or through the courts, whereby the assets are collected and realised, debts paid and balances distributed.
- Dissolution – the breaking up and disposal of a company.

There are different options open to partnerships:

- Partnership administration – a temporary delay of action (moratorium) by creditors to allow a partnership to restructure.
- Partnership Voluntary Arrangement (PVA) – similar to CVA.
- Winding up – liquidation of a partnership whereby the assets are collected and realised, debts paid and balances distributed.

There are different options open to individuals:

- Individual Voluntary Arrangement (IVA) – similar to CVA and PVA.
- Bankruptcy – bankruptcy of an individual by court order.

Let us now consider in more detail receivership and liquidation.

Receivership in England and Wales in relation to companies usually refers to administrative receivership, where an administrative receiver is appointed by a secured creditor, although a receiver may also be appointed by the courts under certain circumstances. The Administrative Receiver has to be an Authorised Insolvency Practitioner. Authorisation is by the Department of Trade and Industry but the actual process of authorisation is generally carried out by the Institute of Chartered Accountants, the Association of Certified Accountants, the Law Society or the Insolvency Practitioners Association. Section 1 of the Insolvency Act 1986 gives the Administrative Receiver powers to take possession of property, raise money, appoint professional consultants and in general carry on the business of the company as the company would had it not been in receivership.

Liquidation is the principal method of dissolving a company and can be:

- Members or shareholders Voluntary Liquidation (MVL) – whereby the members/shareholders appoint an insolvency practitioner to oversee the dissolution of the company, payments to creditors including employees and tax authorities, and repayment of shareholders' capital. This is generally used where it is expected that all creditors will be paid in full. This is commonly used in restructuring companies where a subsidiary is no longer needed or feasible.
- Creditors Voluntary Liquidation (CVL) – where the members and/or creditors pass a resolution that the company cannot continue trading or pay its debts. They usually appoint an insolvency practitioner to act as liquidator. If the members and the creditors each appoint separate liquidators, as is possible, the creditors' liquidator will have precedence.
- Compulsory Liquidation – where the company cannot pay its debts and is forced into liquidation as the result of an application to the court by a creditor or the Secretary of State. The court appoints the Official Receiver[2] as liquidator. The Official Receiver will, if the company has assets to be realised, appoint an insolvency practitioner as the liquidator who will realise the assets, dispose of them and distribute payments to creditors.

In England and Wales, the liquidator takes the place of the directors of the company and only the liquidator or specific persons authorised by the liquidator can bind the company. The liquidator realises the assets of the company and pays the creditors what it can from the funds then available. The liquidator will examine the affairs of the company during the period leading up to the liquidation and will report to the Department of Trade and Industry on its findings.

If it finds that any assets are missing it will search for them and if any have been transferred out of the company in a way that it considers to be not in the best interests of the company it has the power to reverse those transactions. Any monies due to the company, for instance under a construction contract, must be paid by the employer to the liquidator to be distributed as part of the general fund held for the eventual benefit of the company's creditors. Even if the main contract makes provision for the employer to pay sub-contractors direct upon the contractor's default, the employer must not do so once liquidation procedures have commenced.[3] The sub-contractor will, in that case, become one of the company's creditors in the normal way. Failure by the employer to follow this basic principle could result in it paying the same debt twice, once to the sub-contractor and once to the liquidator.

The Insolvency Act of 1986 of England and Wales states that creditors can be:

- secured, fixed charge[4] (such as a bank, mortgage company, finance house etc.);
- reservation of title (creditors can reclaim assets which have not been paid for where full payment was a condition of transference of title);
- debts subject to lien (right to retain possession of assets until settlement of debt);
- preferential unsecured debts (e.g. PAYE, VAT, Customs and Excise, Inland Revenue etc.);
- floating charge;[5]
- non-preferential unsecured (ordinary creditors not falling into any of the above categories).

The distribution of any funds available after assets have been realised and disposed of will be carried out in strict precedence, with fixed charge holders being paid first followed by preferential creditors, floating charge holders and then the remainder of the creditors, who will all be unsecured creditors. Shareholders will be the very last to receive anything and then only after everybody else has been paid in full.

In relation to other countries the Common law approach of England and Wales is considered to be very pro-creditor in its approach to insolvency whereas France is considered the least pro-creditor country followed closely by the US. Whereas under the Common law approach in England and Wales there is no stay on assets, secured creditors are required to be paid first and management is distanced from the reorganisation process, the US allows an automatic stay on assets, has no restrictions for initiating reorganisation and allows the management board to remain in control during reorganisation. France is similar to the US but in addition does not require secured creditors to be paid first as is the case in England and Wales and the US. Germany is closer to England and Wales in its pro-creditor approach except that it does not distance management from the reorganisation process.

Bankruptcy jurisdiction under US law is intended to permit similarly situated creditors, regardless of where they are located, to be treated equally in a bankruptcy or reorganisation case. Discrimination on the basis of citizenship is not permitted. All creditors are given the opportunity to file claims against the State and their recovery is not limited to the assets in their own country. The main distinction between England/Wales and US procedures is that in the US the debtor company remains in possession, while in England and Wales there is an independent administrator.

We often hear of major companies being placed under 'Chapter 11' protection in the US. Chapter 11 is a section of the US Bankruptcy Code which provides a procedure for placing a company which has financial problems under court protection so that the company can attempt to reorganise its finances and make provision for paying its debts, all the while carrying on business in a more or less normal fashion. The company becomes subject to the supervision of the US Bankruptcy Court, a division of the US District Court, and its management continues to operate the business as a 'debtor in possession' unless the Bankruptcy Court appoints a trustee (a trustee is generally only appointed if fraud, dishonesty or gross mismanagement can be shown).

The debtor company has 120 days following filing for Chapter 11 protection within which to propose a plan of reorganisation prepared in consultation with creditors and other interested parties, although this time scale may be extended. If it fails to submit a plan within this period the creditors may propose a plan to the other creditors and interested parties, and the finally agreed plan may be confirmed by the Bankruptcy Court.

Problems arise when the debtor and creditor are located in different countries (cross border insolvency). Where there is a liquidation the general principle is to ascertain what is the corporate domicile[6] of the company in liquidation and to let the court of the country of corporate domicile act as the principal court to govern the liquidation. The other courts would then act as ancillary, as far as they can, to the principal liquidation.

One pertinent comment made in England/Wales is that it is quite consistent with the comity[7] of nations that it should be a rule of international law that if the court finds there is a process of universal distribution of a bankrupt's effects already pending, the court should not allow steps to be taken in its territory which would interfere with that process of universal distribution.[8] In this respect, the US Chapter 11 procedure has been described as not a process of universal distribution but a process of preferential distribution.[9]

In England and Wales the basic principle in cross border insolvency situations is that the law of the place where the liquidation is opened (lex fori) will govern all matters relating to the winding up, including the avoidance of preferences and other transactions. The courts in England and Wales will, however, recognise the ancillary status of a domestic winding up order if a foreign company is also subject to a more significant liquidation in a foreign jurisdiction. It has also been noted that the proper approach by a court in England and Wales would be for it to regard the country of incorporation as the principal forum for controlling the winding up of a foreign company and that, where that company has assets in England or Wales, ancillary winding up procedures should be initiated according to bankruptcy procedures of England and Wales while working in harmony with the foreign courts.

This is a most complicated issue not adequately covered by case law and each case would need to be considered in consultation with an expert in the field taking account of the specific insolvency laws, regulations and customs in the countries concerned.

1. An administrative receiver is an Authorised Insolvency Practitioner appointed by a secured creditor, such as a bank or other debenture holder, and has a principal obligation to his appointer. It is not accountable to other creditors for the way in which it manages the administration in the way that an ordinary receiver would be.
2. Civil servant employed by Department of Trade and Industry in England and Wales – the Official Receiver has Official Receiver regional offices covering court-ordered bankruptcies and liquidations throughout England and Wales.
3. *B. Mullen and Sons (Contractors) Ltd* v *Ross and Another* (1995).
4. Fixed charge – a form of security granted over specific assets which prevents the debtor from dealing with those assets without prior consent of the secured creditor.
5. Floating charge – a form of security granted to a creditor over changing assets of the company (such as stock) – unlike under a fixed charge, the debtor can deal with those assets in its business until it is in default of a payment term.
6. Corporate Domicile – the corporation/company's legal home, usually where its headquarters are located.
7. Comity – the recognition, courtesy and respect which one nation allows within its territory to the legislative, executive or judicial acts of another nation, having due regard both to international duty and convenience, and to the rights of its own citizens or of other persons who are under the protection of its laws.
8. *Galbraith* v *Grimshaw and Baxter* [1910] 1 KB 339.
9. *Mobil Sales and Supply Corp.* v *Pacific Bear (owners)* [1979] Hong Kong.

Insurance Claims

As with insolvency, there are numerous specialist books concerning insurance written by experts in the field and I would not presume to be such an expert. In this section we will consider only our involvement in settling insurance claims (an insurance claim in this context meaning a claim by the contractor against an insurance policy for which the contractor is either the sole or joint policy holder) since this is the principal area of insurance that we will be most concerned with in the administration of our contract.

When an insurance claim has been identified the insurers must be notified using standard reporting procedures and forms that should have been agreed at the outset of the contract works. The formal notification needs to be a brief summary of events sufficient to log the claim in the insurer's system at the earliest opportunity. The subsequently-issued full report should be prepared in close consultation with all personnel actually involved with the incident or closely associated with it and this should be prepared while memories are fresh and clear.

Although this initial report will need to be modified at a later date to accommodate facts discovered after the event, the basic facts and observations uttered by those actually on the scene should not be changed as to do so could seem to the insurance company as a deliberate attempt to distort the facts to enhance the claim. We will need to liaise closely with the person responsible for preparing the incident report/insurance claim and other concerned staff to ensure that:

- the insurer is only informed of the precise and unquestionable facts of the incident;
- no assumptions are made regarding the incident, its causes, its effects on the works or the remedial/replacement works necessary;
- there is to be no acknowledgement or acceptance of any liability or implication or inference to this effect;
- estimates are to be prepared as accurately as possible in the knowledge that these will form the basis for any settlement figure which the insurers will eventually propose. At this stage it is preferable to be over-generous in estimation rather than to try to restrict the cost to the bare minimum since at the end of the day there is every likelihood that additional works will need to be carried out which were not originally anticipated and which the insurer would then be loath to add on to the budget which it will have allocated after its initial review of the claim.

All site staff, including sub-contract personnel, must be clearly informed about the incident and directed to keep individual records of all work affected by or carried out as a result of, the incident until the insurance claim has been settled. Only designated staff should have any dealings with the insurance company's representatives and assessors since to do otherwise would run the risk of confusing them with different people's recollections and theories, not precise and unquestionable facts, and this in turn would inevitably have a delaying effect on the insurance settlement.

The staff must be left under no illusion that the insurance company's surveyor or loss adjuster is there to reach a fair assessment of the claim. In fact they are there to minimise the insurance company's exposure, and external loss adjusters are very often engaged on a sliding scale fee basis, i.e. the more they reduce the claim, the higher their fee will be.

The person responsible for coordinating the insurance claim will immediately establish a Claim Recovery file that must contain as a minimum:

- incident reports;
- witness statements;
- photographs;
- inspection records;
- works necessarily carried out resulting from and caused by the incident;
- consumables expended in carrying out such works;
- manpower, plant, equipment and material records from the date of the incident to the end of the repair or replacement, including completion of any tests subsequently required;
- daily progress reports from the date of the incident to the end of the repair/replacement and tests.

The insurer will send to the scene of the incident either its in-house surveyor or, more probably, a representative from its local loss adjuster. The contractor shall give the insurer's surveyor or loss adjuster all necessary assistance and cooperation sufficient for it to be able to:

- identify and survey the damage;
- collect all available information relating to the incident from which the insurance claim arises, including the circumstances of the incident and any steps taken to mitigate the damage caused;
- review all documents, whether at the site or elsewhere, relating in any way to the incident including but not being limited to site diaries, daily progress reports, inspection records and reports, photographs taken prior to, at the time of and subsequent to the incident etc.;
- verify estimates regarding the costs of labour, materials and plant/equipment necessary to remedy the effects of the incident and any time or cost effects which the incident and the subsequent wait for remedial works to be completed will have on the overall construction schedule.

We should be involved in all the commercial discussions regarding the incident and should ensure that the strategies agreed upon at the outset are followed through without deviation in order that the insurer is given the level of confidence in our operation that it needs to have if our claim is to be met. As soon as the insurer has agreed the claim the contractor can proceed with the remedial works. The insurer's confirmation of acceptance must be in writing, unless the contract and policy details state otherwise, and the contractor must process this immediately it is received.

It is extremely likely that the insurer will not make an instant decision except in the most straightforward of cases. In all probability it will review its report and make a qualified decision pending further works being carried out to determine both the cause and the overall effect that this might have had on the remainder of the project. If the contractor is either instructed or permitted to carry on with a limited amount of repair or replacement works by the insurance company pending final settlement, we must ensure that there is agreement on costs and terms of payment for these ad hoc works before they are initiated. Accurate contemporary records must be kept of all such works and of all other works affected by the incident that gave rise to the claim. The insurer will eventually make a final decision on costs, which will rarely satisfy the contractor or its sub-contractors but which will in all probability be the best they can hope to achieve.

The emphasis, which cannot be overstated, is on keeping full and accurate contemporaneous records throughout the whole period commencing with the incident itself and ending on the completion of remedial works and settlement of the claim.

Interrogatories

Interrogatories are a set of written questions sent by one attorney to the other attorney for its client to answer under oath or under penalty of perjury as part of the pre-hearing discovery process. For the most part, any question is fair game, so long as it is relevant to the case. The written answers must be provided within a designated time period, usually within 30 days.

The party answering the interrogatory is obliged to sign the answers confirming that they are complete and truthful to the best of its knowledge. If the party is not a natural person then the party's attorney or other representative may sign the answers, in which case they need not be verified, although the party propounding the interrogatories may request that the answers be verified by a person employed by the party rather than merely representing it as an attorney would, and having the most knowledge of the subject matter covered by the interrogatories.

The system described above is used to a considerable extent in the US where several states ask basic 'form' interrogatories comprising set questions on a printed form, with an allowance for 'supplemental' interrogatories specifically relevant to the lawsuit. A basic 'form interrogatory' could include such questions as 'state the name, position, company, address and telephone number of each person who witnessed the incident,' and 'state the amount that you admit you owe and describe and explain in detail your basis for calculating that amount'. The supplemental interrogatories accompanying the basic form would then ask questions specific to the particular subject matter in dispute.

In England and Wales a party to an action is required to apply to a Master[1] for directions for an order permitting him to serve interrogatories on an adverse party. The other party would then reply on an affidavit.[2]

In the US there is also another class of interrogatories called 'special interrogatories', which are written questions asked by a judge to a jury to establish whether the jury's answers are consistent with their verdict. Various US State Codes of Civil Procedure provide for special interrogatories, one in particular[3] stating that:

> *The jury may be required by the court, and must be required on request of any party, to find specially upon any material question or questions of fact submitted to the jury in writing. Special interrogatories shall be tendered, objected to, ruled upon and submitted to the jury as in the case of instructions. Submitting or refusing to submit a question of fact to the jury may be reviewed on appeal, as a ruling on a question of law. When the special finding of fact is inconsistent with the general verdict, the former controls the latter and the court may enter judgment accordingly.*

The purpose of a special interrogatory is to test the general verdict,[4] and to have proper effect it must relate to an issue of ultimate fact, otherwise, there is no point in having the jury answer.

Interrogatories differ from depositions in that interrogatories will be in formal written documents prepared by the party's attorney with the party's help. In this way the attorney has more control over how the party's responses are delivered than it would have if the party were to be asked the same question in a deposition. The answering party would benefit from help from its attorney in understanding the meaning (sometimes hidden) of the questions and to avoid adding wording in its answers that could be interpreted against the answering party.

Attorneys, particularly in the US, often favour the use of depositions rather than interrogatories when they are searching for certain specific types of information from the opposing attorney's client in order to minimise the degree of assistance and interference by the opposing party's attorney in their compilation. Objections as to the relevancy or clarity of the questions or answers may be raised either at the time the interrogatories are answered or when they are used in trial.

Similar to depositions, interrogatories can be used in a hearing to discredit a witness, where the witness's answer given at the hearing differs from that given to the same or similar question in an interrogatory. This is a classic way of putting a party's credibility on the line.

Generally interrogatories are thought of as a means to expedite the discovery process. If answered promptly and completely they can act as the foundation for additional discovery by depositions or other means open to the attorneys involved in the case. If, however, the party served with the interrogatories fails to answer either promptly or at all, this can serve to delay the whole discovery process and can bring about collateral disputes that will then need the intervention of the tribunal or court to resolve.

When it comes to effecting service of interrogatories on a foreign national the Hague Convention[5] on the collection of evidence requires specific documents and procedures to be used, as well as specific rules of compliance to be followed.

1. In England and Wales the jurisdiction of a Master is that of a Judge in Chambers (with some exceptions).
2. An 'affidavit' is a written statement by a party, referred to as the deponent. The statement is made under oath, or affirmed, in front of a person permitted by law to administer an oath and must relate only to statements which the deponent has knowledge of. In special cases it may contain statements of information and beliefs and must then state the sources of such information and beliefs.
3. State of Illinois Code of Civil Procedure at 735 ILCS 5/2-1108 (ILCS refers to Illinois Compiled Statutes) – other US states have similar statutes in place.
4. *Simmons* v *Garces* (2002) 198 Ill. (2d) 541; 763 NE (2d) 720.
5. Hague Convention on the Collection of Evidence Abroad in Civil and Commercial Matters.

K

Kick-Off Meeting

As soon as the contract has been signed and a commencement date agreed it is advisable to hold a 'kick-off' meeting at which the main and sub-contract personnel to be directly involved in the execution of the contract can have the opportunity to meet and discuss procedures to be adopted, outstanding information required by each side and any other matters which need discussing prior to commencement of work. This meeting should establish lines of communication and be a further step in the building of good relationships between the contractor and his sub-contractors.

A similar kick-off meeting should be held between the contractor and the employer and the employer's representatives, or alternatively a combined meeting may be held including the whole project team. This would depend upon the size of the project and the number of parties involved – too large a meeting would only be counterproductive. The basic principles discussed in this section can be applied to the kick-off meeting between the employer and the contractor.

There follows a suggested agenda for a kick-off meeting between the contractor and its sub-contractor. Each party should ensure that this meeting is attended by its:

- project manager;
- site manager;
- contract/commercial manager;
- planner;
- cost controller/quantity surveyor.

The purpose of the sub-contractor kick-off meeting is to:

- identify and establish the working interfaces between the contractor, major sub-contractors and others working on or in connection with the site and the works;
- explain site coordination procedures, rules and regulations, with particular regard to health and safety rules and regulations;
- establish a time frame for mobilisation of the sub-contractor's site facilities, labour, plant and equipment;
- ensure that there is no confusion between the contractor and the sub-contractor regarding any issues related to the sub-contract works;
- exchange documents and information.

It is important that any meeting should have an agenda which is circulated to the parties who will be attending the meeting well in advance of the meeting in order that they can prepare information to be given out at the meeting, questions which they will need to ask at the meeting, and answers for questions which they will be asked at the meeting. This is particularly important with respect to a kick-off meeting which might well be the first contact between many of the project team members and which will set the tone for the management of the project. The meeting should be minuted and will serve as a reference point should any disputes arise between the parties later in the term of the contract.

Proposed kick-off meeting agenda[1]

Documents and information to be provided by sub-contractor to contractor:

- parent company Bond or Guarantee;
- Performance Bond;
- names and addresses to whom and to which contractual notices are to be served;
- copies of policy documents in respect of insurances required under the sub-contract, including certificates of insurance, insurance schedules and receipts confirming payment of premium;
- names and titles of sub-contractor's management and administrative staff to be engaged on the project with their duties and line management responsibilities;
- programme for the sub-contract works for approval;
- method statement in respect of the sub-contract works for approval;
- lines of communication;
- cash-flow forecast;
- confirmation of receipt of all necessary drawings and other particulars to enable sub-contract works to commence;
- confirmation that all statutory and public utility service provider approvals and licences etc. have been obtained;
- confirmation that labour, plant and materials have been arranged and are ready for mobilisation on the due date;
- details of any sub-letting arrangements contemplated by sub-contractor;
- confirmation that all sub-contractors' sub-sub-contractors and suppliers are aware of the programme requirements and are confident that these can be achieved;
- information outstanding and dates by which it is required in order that adherence to the programme is not jeopardised;
- samples of materials to be used (if appropriate).

Documents and information to be provided by contractor to sub-contractor:

- names and addresses to whom and to which contractual notices are to be served;
- documents or information of any kind which the contractor is obliged to provide to the sub-contractor and which is outstanding at the time of the kick-off meeting;
- names and titles of contractor's management and administrative staff to be engaged on the project with their duties and line management responsibilities;
- evidence of the contractor's compliance with the insurance requirements of the contract;
- communication and coordination procedures;
- site access and entry permit procedures;
- details of on-site and off-site parking facilities;

- details of site health and safety regulations and procedures, works and local regulations;
- location of site benchmarks;
- location of terminal points (if applicable);
- details and location of service mains connections for site power, gas, water, drainage etc.

Matters to be agreed:

- date of commencement of sub-contract works;
- location of sub-contractor's site offices, toilets, canteens facilities, material storage buildings/warehouses, lay-down areas and sign boards;
- sub-contractor's precise requirements as to location of service outlets, connections and required capacities, which contractor has to provide under the contract (water, gas, electricity, drainage etc.);
- sub-contractor's site security obligations and requirements;
- payment procedures (format of applications required by the contractor and the dates by which contractor must receive payment application each month);
- daywork procedures, including method and timing of necessary authorisations, rates to be applied (if not already included in sub-contract agreement or purchase order) and format of Daywork Record Sheet;
- dates and format for project site meetings which the sub-contractor must be available to attend if so required at any time;
- dates and times of sub-contractor review meetings which the sub-contractor must attend.

1. A similar format could be used for the employer/contractor kick-off meeting adapted as necessary to reflect the different status of the parties and the particular nature of the contract.

Latent Defects – Limitation of Liability

It is critical in the management of construction contracts that we are aware at all times of the limitation period within which all actions must be commenced. These periods vary under the specific legislations enacted by each country and this discussion will in the main consider the situation under legislation in England and Wales, although references are made to the different periods applicable in several other jurisdictions. The same or similar principles to those discussed in this section will apply under most other jurisdictions utilising the specific limitation periods enacted by their own legislation.

The claimant has a specific period under relevant legislation within which it is obliged to submit its claim or serve notice of arbitration. This period commences in England and Wales on the date on which the claimant's cause of action arose but the commencement date can vary in other jurisdictions.

Construction contracts will either be executed as a deed (previously better known as 'under seal') or under hand (a simple contract). In England and Wales the former has a limitation period of 12 years whereas the latter has a limitation period of just 6 years.[1] Limitation periods in other countries vary; for instance under German law they can either be 3 or 30 years,[2] Canada between 1 and 6 years subject to jurisdiction, US between 4 and 10 years subject to jurisdiction,[3] South Africa 3 years and South Korea between 1 and 5 years. These periods might have been reviewed by the time that you have concerns in respect of a particular contract and it is therefore worth checking the limitation periods and restrictions in whichever country you intend operating in as part of the pre-contract risk assessment exercise so that you are absolutely sure of the current situation.

Some countries including Australia, Belgium, Egypt, France, Italy, Luxembourg, UAE and Venezuela have a decennial liability law that requires contractors to take out decennial liability insurance in respect of building defects that occur within ten years of completion.[4] This applies only to buildings, with their equivalent of the Latent Damage Act in England and Wales applying to everything else.

In Common law jurisdictions the limitation periods start to run from a point known as the 'arising of a cause of action'. This in itself is not a straightforward term and is often misunderstood. A cause of action is the summation of relevant facts that the claimant believes justify its case. In contract, this is a called breach of contract and the time starts to run when a breach occurs which cannot be remedied without causing the claimant damage or loss. In the case of a claim for delay, for instance, the time will start to run when the employer has had the last opportunity to review the claim and has either chosen to ignore or to reject it. From the employer's point of view, if there are defects that the contractor has not remedied, the limitation period for any claim the employer intends making against the contractor will start on the date of practical/substantial completion of the works.

The beginning of limitation periods under Civil Codes differs from that under Common law. For instance, under the German Civil Code:

(1) the standard three years limitation period begins[5] upon the expiry of the year in which:

- the claim has arisen, and
- the obligee becomes aware of the circumstances giving rise to the claim and of the identity of the obligor or ought to have become aware of those matters but for its gross negligence;

(2) the limitation period for established claims begins[6] on the date when:

- the decision becomes final and absolute; or
- the enforceable title is created; or
- the claim is established in the insolvency proceedings

but not before the claim arises.

In a negligence claim in tort, where the limitation period in England and Wales is 6 years (subject to a qualification brought about by the Latent Damage Act,[7] as noted in the next paragraph), the cause of action is the breach of a duty of care resulting in loss or damage and the time does not start to run until this loss or damage has actually occurred.

As a result of the different causation start dates under contract and under tort, a claimant who has missed the boat for initiating a claim under contract sometimes is able to claim under tort where the damage occurs at a later date than that which would have been the latest date for a claim under the contract to have been possible. The Latent Damage Act of England and Wales introduced a qualification to the period of limitation applicable for a claim for damages for negligence (which could include latent defects) in order to overcome the possibility that a situation causing loss or damage had occurred without the claimant being immediately aware of it. The Act introduced a long stop provision stating that a claim for damages for negligence under tort could be submitted within a period of 15 years from the accrual of the cause of action provided that it was brought within three years from the date of discoverability, i.e. the date of the claimant becoming aware:

- that the relevant damage was sufficiently serious to take court proceedings;
- of the facts about the damage caused by the defendant;
- of the defendant's identity; and
- of the name of the person who committed the negligent act, if other than the defendant, and of the reasons for bringing the action against the defendant

The starting date is only reached when knowledge of all the factors listed above is available to the claimant and it realises it has a cause of action, or on the date when it could reasonably have realised it.

'Knowledge' in this context includes the actual knowledge of the claimant and also knowledge that the claimant should reasonably have been able to obtain either on its own account or by seeking appropriate expert advice where it would have been reasonable for it to do so. If a negligent act was committed more than 15 years before the issuance of court proceedings, the court could not consider the claim even if the actual damage had only occurred within the last six years. If the claim was related to a property purchased by the claimant and if that property was already in a damaged condition at the time of purchase, the three-year period would commence from the date of the previous owner's first knowledge of that damage.

1. To be found in the Limitation Act 1980 of England and Wales amended by the Latent Damage Act 1986 and the Consumer Protection Act 1987.
2. See German Civil Code Articles 195 (three years' standard limitation period) and 197 (ten years in specified situations, including in the case of:
   (i) claims that have been declared final and absolute;
   (ii) claims based upon enforceable settlements or enforceable documents; and
   (iii) claims that have become enforceable by virtue of having been established in insolvency proceedings).
3. Some US states have different limitation periods in respect of oral and written contracts, for example California's limitation period is two years for oral compared to four for written, Connecticut and Washington three for oral compared to six for written, and Georgia four for oral compared to twenty for written.
4. See separate section on 'Decennial Liability' on p.84.
5. German Civil Code Article 199 – this also sets out the time bar restrictions applying to all claims.
6. German Civil Code Article 201.
7. Latent Damage Act 1986 of England and Wales.

Letters of Credit

In the simplest terms, a letter of credit is a document issued by the bank of the buyer of goods or services to the seller of those goods or services promising that it will pay a specified amount to the seller upon satisfaction of certain preconditions (usually the presentation by the seller to the buyer's bank of specified documents – documentary letter of credit), usually within a specified time. Letters of credit normally describe the goods or services concerned, the documents required, a shipping date and a date after which the payment will no longer be made. The letter of credit is a separate contract and as such it is independent from the sales contract it relates to.

A letter of credit gives the seller comfort knowing that it will receive money for its supplies even if the buyer changes its mind or becomes insolvent – the bank is still obliged to pay under such circumstances. No worldly-wise contractor would consider commencing work on an international contract without the appropriate letters of credit being in place guaranteeing payment of its approved invoices.

The buyer, on the other hand, has the comfort of knowing that it will not have to pay until the required documents are produced by the seller and the bank has verified that they are consistent with the conditions indicated in the letter of credit, which in turn means that the goods have been delivered to the right place within the specified time, or that the employer's representative has certified that the work invoiced has in fact been satisfactorily carried out in accordance with the contract.

The buyer can usually negotiate more favourable terms with the seller when it agrees to pay by letter of credit because of the reduction in the seller's risk. The buyer also has the advantage of delaying payment until the letter of credit conditions have been met, which helps its cash flow situation. The payment terms of a letter of credit can be at sight, at a specific date some time after sight, or a combination of the two. Payment terms stipulated in a letter of credit should be the same as stipulated in the sales contract.

The conditions of the letter of credit are determined by the buyer, who instructs its bank accordingly. Banks only verify that the documents are in order and take no responsibility or liability for the goods or services provided in relation to the documents.

There are several types of letter of credit and these will now be discussed briefly:

- ***Irrevocable*[1] *letters of credit*** – the opening bank assumes an obligation to pay upon satisfaction of the letter of credit conditions and this obligation cannot be cancelled or amended by any party without the unanimous agreement of all the other parties.
- ***Revocable letters of credit*** – in this case any party to the letter of credit can cancel or amend it without reference to or the agreement of any other party to the letter of credit. This is a major disadvantage to the seller and so is only used under special circumstances.

- *Confirmed letter of credit* – usually used when the seller has doubts about the buyer's bank or its country of origin. In this case both the confirming bank and the opening bank assume the obligation to pay the seller upon satisfaction of the letter of credit conditions.
- *Unconfirmed letter of credit* – this is opposite to a confirmed letter of credit and does not bear the undertaking of any bank other than the issuing bank.
- *Transferable letter of credit* – a transferable letter of credit specifically permits transfer from the original beneficiary to one or more secondary beneficiaries.
- *Stand-by letter of credit* – this covers the situation whereby a buyer has to make a simple transfer of funds to the seller for the goods or services. If it fails to do so (defaults) the seller presents the requisite documents to the bank and payment is made against the stand-by letter of credit. In this case the documents might only be a copy of the unpaid invoice and the seller's written claim to the bank for payment against the stand-by letter of credit.

There are various other terms used in letter of credit dealings that it might be useful to touch on briefly:

- *Accepting bank* – a bank that by signing 'accepted' on a timed draft accepts responsibility to pay when the draft becomes due.
- *Advice of credit* – communication by the advising bank that a letter of credit has been issued.
- *Advising bank* – the advising bank (a 'correspondent bank') acting as an agent of the issuing bank whose primary responsibility is in establishing the authenticity of the credit and supplying details to the beneficiary without any responsibility or engagement in the transaction. Usually located in the seller's country.
- *Confirming bank* – the bank that adds its name to the letter of credit alongside the opening bank's as an added guarantee (see 'Confirmed letter of credit' above). A 'confirming bank' confirms that it will honour a letter of credit issued by another or confirms that such letter of credit will be honoured by the issuing bank or by a third bank.
- *Correspondent bank* – a bank involved in a bank to bank relationship essential in the letter of credit process where two banks exchange signatures, test keys and settlement of funds agreements.
- *Cumulative letter of credit* – a revolving letter of credit that permits any amount not utilised during any of specified periods to be carried over and added to amounts available in subsequent periods.
- *Draft* – a financial instrument (bill of exchange) indicating the debt of one party to another (i.e. buyer to seller in a letter of credit transaction).
- *Evergreen letter of credit* – a letter of credit that automatically renews itself beyond the specified expiration date.
- *Issuing bank (also referred to as the opening bank)* – the bank that opens the letter of credit on behalf of the buyer, giving the primary undertaking to honour the letter of credit.

- *Negotiable letter of credit* – a letter of credit issued in such a way that it allows any bank to negotiate the documents, incorporating the opening bank's engagement and honouring presentation of drafts as long as all the terms of the letter of credit are met.
- *Negotiating bank* – the bank nominated on a letter of credit, usually other than the opening bank, to negotiate the draft (bill of exchange) (i.e. check the documents, pay the seller, seek reimbursement from the opening bank).
- *Opening bank* - see 'issuing bank' above.
- *Paying bank* – the bank designated in the letter of credit as the bank that will honour drafts drawn under the letter of credit upon presentation of the requisite documents.
- *Reimbursing bank* – a bank nominated by the opening bank to pay either a negotiating, accepting or paying bank in a letter of credit transaction. The reimbursing bank is not a party to the letter of credit.
- *Revolving letter of credit* – a letter of credit that includes a provision for reinstating its face value after being drawn upon, within a specified period of time. This is usually used to finance a series of similar purchases over a period of time.

The International Chamber of Commerce's 'Uniform Customs and Practice for Documentary Credits', 1993 Revision effective 1 January 1994 (ICC Publication No. 500) applies to all documentary credits where they are incorporated into the text of the credit. These customs and practices are binding on all the parties unless it is expressly stipulated otherwise in the credit. Bank to bank reimbursement arrangements under Article 19 of the Uniform Customs and Practice publication are supplemented by the ICC's 'Uniform Rules for Bank to Bank Reimbursement',[2] which set out a series of concepts to facilitate the reimbursement process.

1. Under Article 6 of the International Chamber of Commerce's 'Uniform Customs and Practice for Documentary Credits' ICC Publication No. 500 effective 1 January 1994, if nothing is stated on the Letter of Credit it is deemed to be irrevocable.
2. International Chamber of Commerce Publication No. 525; ISBN 92-842-1185-9.

Letters of Intent

A 'Letter of Intent' may also go under the title of: 'Expression of Interest', 'Agreement in Principle', 'Memorandum of Understanding', 'Memorandum of Intent', 'Head of Agreement', 'Protocol', 'Term Sheet' and 'Letter of Awareness'. For all practical purposes, they are all the same things.

A letter of intent may or may not create a contract depending upon the exact wording used and whether it contains the essential terms necessary to obligate two parties to a transaction. A letter of intent is usually intended to be non-binding; however, it is viewed as a binding document if the parties intend it to be binding or if it stays silent on its intended status.

The letter may only amount to a statement of the sender's intention to enter into a contract in the future. Such statements of intent are merely agreements to negotiate and are not recognised or enforceable under English Law[1] and will not generally impose obligations on either party.

If the letter of intent merely states that the tender has been received and contracts are being prepared then it would be extremely unwise to commence any work until either the contract was formalised or a more specific letter of intent was received. If, however, the contractor decides that it would be advantageous for its programme if it were to commence design drawings, for instance, at that stage then it must accept that it is doing so at its own risk.

The biggest problem encountered in the practical application and use of letters of intent comes when payment is due. It is imperative that the letter of intent either defines exactly what the scope of work it covers entails and the price that will be paid for its execution (for instance, design drawings to be undertaken up to a certain stage for a set fee) or defines exactly how work carried out under the letter of intent will be valued should the contract not proceed (for example, site preparation/substructure works to be valued at rates included in the accepted tender bills of quantities). A letter of intent can be a useful means of achieving a timely start to a project when administrative difficulties are delaying the official contract commencement. Letters of intent do, however, have to be treated with care and certain precautions need to be taken to formalise and preserve the intent of the parties.

If work was carried out under a letter of intent which specifically requested or instructed such works to be commenced but which was incomplete or silent as to the terms under which the works were to be performed and did not specify the procedures for calculating and making payments, here there would be no contract and all work legitimately carried out would be valued under the principles of quasi contract, restitution or quantum meruit valuation.[2] In addition:

- there will be no express or implied warranties as to fitness for purpose, quality, delivery or title;
- there will be no right of set off or counter claim because there will be no contract and so no breach of contract;

- the contractor will be under no obligation to complete the works and can cease work at its pleasure;
- the parties will not have the right to adjudicate under the Housing Grants, Construction and Regeneration Act 1996 since there will be no agreement under the Act for the carrying out of construction operations.

Where a letter of intent refers to specific contract terms applying then those terms will apply even though the actual contract has not been entered into (i.e. they will be treated as incorporated by reference). The letter of intent could be so broad as to even include all the terms of the intended contract. The contract will in any case apply retrospectively to the date of commencement of works under the letter of intent once the contract is signed.

If the parties do not want to be legally bound by a letter of intent, they should be very careful in their choice of words and title, should not charge non-refundable fees and should ask their lawyer to draft the letter of intent for them. A non-binding letter of intent should contain:

- a disclaimer stating that it is not intended to be, and does not constitute in any way, a binding or legal agreement between the parties (there is a presumption that a letter of intent that states that it is not intended to be legally binding will not be legally binding);
- a statement that the parties will attempt to negotiate a binding contract but that if such a contract is not negotiated to conclusion within a certain period of time then the letter of intent is void. In Common law jurisdictions 'subject to contract' clauses will usually rebut the presumption of contractual intent and indicate that the parties are still negotiating and the terms have not yet been finalised.[3] 'Subject to contract' is, however, not a recognised term in Civil law jurisdictions, so it should not be relied upon when dealing with foreign parties (subject to appropriate legal advice from a lawyer familiar with the jurisdiction concerned);
- a statement that the letter of intent does not include all the essential terms required in order for it to be construed as a contract between the parties, setting out those essential terms that are missing;
- a provision that either party can walk away from the transaction at any time, with or without cause;
- a provision, if there is to be an exchange of confidential information at this initial stage, that a separate confidentiality agreement be entered into which will impose binding obligations of confidentiality upon the parties that will survive the non-binding letter of intent;
- a provision that the parties are under no obligation to use their best efforts, or to act in good faith, to negotiate a contract.

In explanation of the last bullet point, a duty of good faith in contractual negotiations is not generally recognised in English law. As the law currently stands in England and Wales, even where parties explicitly agree to negotiate in good faith, this is not an agreement that will generally be enforceable as the courts are reluctant to restrict the freedom of parties to negotiate. In most other international jurisdictions based upon Civil law principles, however, a duty of good faith is recognised, for instance:

- The French Civil Code states that:'They [*Agreements*] must be performed in good faith.'[4]
- The German Civil Code states that: 'The obligor must perform in a manner consistent with good faith taking into account accepted practice.'[5]

A duty of good faith in contracts governed by the law of these countries will create obligations requiring each party to:

- inform the other party of all of the relevant points that it could not otherwise discover;
- observe reasonable diligence in the performance of pre-contractual and contractual obligations;
- observe moral and ethical standards of behaviour.

If the parties intend the letter of intent to be binding it should state in clear, unequivocal and express terms:

- that the parties intend to be immediately bound by this letter of intent;
- that its enforceability is not contingent upon the parties entering into a formal agreement;
- the scope of work it covers and what this entails in detail;
- the price that will be paid for its execution and the method/timing/frequency of payments;
- exactly how work carried out under the letter of intent will be valued should the contract not proceed.

The very fact that a letter accepting a quotation states that it is a letter of intent does not necessarily mean that a contract has been formed. The fact that non-critical elements of the agreement have not been agreed upon at the time of issuance of the letter of intent will not necessarily mean that a contract has not been formed. The courts will review the nature of the contract envisaged by the parties and the relative importance of those elements mentioned above when making the decision as to whether or not a contract has in fact been formed by the issuing of the letter of intent.[6] In determining whether the letter of intent does in fact constitute a binding contract it has been noted in court that the following key ingredients of a binding contract must be in place:[7]

- both parties must appear to have intended to make a contract;

- there must be agreement on all the principal relevant terms of the contract with sufficient certainty for it to be commercially viable;
- there must be a clear offer and acceptance; and
- there must be consideration.[8]

The issue of whether or not the letter of intent is binding or non-binding is a complex one and has given rise to many interesting court decisions. The fact that the letter of intent contemplates the execution of a formal agreement is not necessarily conclusive of its non-binding character. Parker J is quoted as stating:[9]

> *It appears to be well settled by the authorities that if the documents or letters relied on as constituting a contract contemplate the execution of a further contract between the parties, it is a question of construction whether the execution of the further contract is a condition or term of the bargain or whether it is a mere expression of the desire of the parties as to the manner in which the transaction already agreed to will in fact go through. In the former case there is no enforceable contract either because the condition is unfulfilled or because the law does not recognise a contract to contract. In the latter case there is a binding contract and the reference to the more formal document may be ignored.*

Courts have enforced the terms of a letter of intent that stated that it was subject to the approval of the board of directors of one of the parties, even though it was disclosed that the board's approval was never actually sought. Courts have also held that the parties had acted in reliance on the letter of intent as if it were enforceable and, by those actions, had made it enforceable.

In one case the letter of intent stated that it:

> *is not intended to create, nor do you or we presently have, any binding legal obligation whatever . . . however, it is our intention, and we understand, your intention immediately to proceed in good faith in the negotiation of such binding definitive agreement.*

The US courts held in this case that the word 'however' coupled with the words 'proceed in good faith' overrode the disclaimer that the letter of intent was non-binding and created an obligation that it was in fact binding.[10]

The moral is to keep any letter of intent as brief as possible, including just enough details to get the other party started and no more.

1. *Courtney & Fairburn Limited* v *Tolaini Brothers (Hotels) Limited* (1975) 1 All ER 716.
2. *OTM Limited* v *Hydronautics* (1981).
3. *Chillingworth* v *Esche* (1924) 1 Ch. 97 – It was held that it would be a 'strong and exceptional case' where the use of 'subject to contract' fails to rebut the presumption of contractual intent.
4. French Civil Code, Article 1134.
5. German Civil Code – Law of Obligations, Article 242.

6. *Twintec Limited* v *GSE Building and Civil Engineering Limited* (2003) EWHC 605 (TCC) 24 March 2003.
7. *Trollope and Colls* v *Atomic Power Constructions Limited* (1963) 1 WLR 333.
8. See 'Consideration' section on p.57.
9. *Von Hatzfeld-Wildenburg* v *Alexander* (1912) 1 Ch. 284.
10. *Schwanbeck* v *Federal Mogul Corp.* 412 Mass. 703 (1982).

M

Minutes of Meetings

When the contract works have been virtually completed and the outstanding commercial work comes down to resolving outstanding disputes, that is the point when we realise the worth of all those boring hours we spent taking notes at and writing up minutes of a wide range of site meetings.

Minutes are an accurate record of the discussions that took place between various parties to and personnel involved in the contract either pre- or post-contract. In order to ensure that they are an accurate record of the actual discussions it is advisable to insist upon writing the minutes ourselves or for one of our assistants to do so under our control. In this way we can ensure that nothing of any importance is missed and we can also, to the extent that it is fair, reasonable and ethical to do so, select to some degree the relative importance to be given to the various topics.

Both the employer and the contractor will benefit from an element of selective recording, skimming over anything which is unrelated to the project works or which unnecessarily labours a point that has already been discussed and agreed. It can happen that the writer tends to put a slight emphasis on the points that are good for its side of the contract team and less emphasis on those which are not, but such an approach must be entertained with caution. In any case, the minutes will be submitted to the other parties for review, correction and agreement prior to appending their signatures so any unfair manipulation should not pass the scrutiny that takes place at that time.

Under no circumstances should statements be recorded in minutes unless those statements were actually made at the meeting concerned (unless they are qualified to that effect and have the other party's agreement to their inclusion) – minutes are *not* to be used as a means of including arguments or statements which their proponent forgot to make at the meeting or which arose after the meeting had taken place.

Meetings held between the tender submission date and the date the contract was signed are very often appended to, and intended to form part of, the contract. In such a case it is extremely important to ensure that there is nothing included in the meeting minutes that goes against the intention of the contract. Parties to such meetings should approach them with an element of sensible caution. It is not unknown for a contractor to state in the exuberance of a pre-contract meeting that it is confident that nothing could prevent the project being completed on budget and within the contract period.

Such a statement is foolish and will be likely to come back to haunt its author at a later stage when problems are encountered and work delayed at extra cost to the owner as I witnessed on a project in Saudi Arabia, a country where virtually nothing is completed within budget or within the contract period. Needless to say the Saudi client hung on to this statement every time it delayed in approving drawings, made significant changes to the works, including changing the external façade from copper cladding to Riyadh stone (when the copper cladding was well under way), added extra works and so on.

It is unwise to make such a statement or promise at any time, particularly at a contract meeting, unless that statement or promise has previously been discussed amongst the project team and decided upon as the team's adopted policy or strategy. If the meeting is of a specialist nature only those with the specialist knowledge to back up their statements should be allowed to put forward opinions.

On occasions meetings will be held on a 'without prejudice' basis and if this is the case then the meeting minutes should also be noted as being 'without prejudice'.[1] On other occasions meetings might just be informal discussions whilst having a cup of coffee or walking around the site. In this case there would not be official written minutes; however, it is always wise to record such meeting discussions either in a file note or as a diary entry. Should there be a dispute at a later date even those notes made at the time of the discussion would have considerable weight in front of the tribunal hearing the dispute.

Wherever possible it is advisable to have a formal agenda stating the subjects to be discussed at a meeting. If the meeting is a regular project progress meeting then the minutes of the previous meeting will usually act as the agenda and everyone will know what subjects will be discussed. In such a case it is normal to discuss matters arising from the previous meeting either as the first topic after acceptance of the previous minutes, or possibly for them to be discussed as each agenda item is reached. New items can be introduced under the heading of 'Any Other Business' if they are one-off items or as additional agenda items if they are likely to occur at a series of meetings. One-off meetings will definitely need agenda as without a formal structure of some kind it is extremely difficult to exercise control over any meeting.

It is wise for each party to nominate a spokesman, either informally amongst themselves or, if deemed appropriate, formally in front of the meeting, and for that person to be the principal or sole speaker on a particular subject. If too many representatives speak there is that much more chance of something being said that will eventually be either misunderstood or used against the offering party. If in doubt, as in most situations, it is best to say nothing. If, however, there is a strong reason to disagree with what one party has put to the meeting then it can be equally dangerous to say nothing, this being open for interpretation as tacit acceptance of what was stated.

The last and most important rule to be applied to minutes of meetings is to check what has been written before signing the minutes as a true record. It is advisable to have the minutes ready for signature as soon as possible after the meeting has closed. It is not unusual to see handwritten minutes passed around for acceptance immediately the meeting has closed but in these days of 'lap tops' and 'palm tops' it is far better to have typewritten copies printed for everybody to read clearly. Unless the meeting was an extremely long one, which again is unadvisable, the minutes should be ready in draft form within a matter of hours for the participants to review and comment on.

Review comments made by the participants should be listened to and, if valid, the minutes should be amended accordingly prior to signature by the parties. The checking process is not, however, an opportunity for a participant to retract or modify any statement which it made at the meeting. As mentioned earlier, meeting minutes are intended to be accurate records of discussions and, if a participant believes that it, or another party at the meeting, made an inappropriate or incorrect statement at the meeting or that its or the other party's words were misconstrued or misrepresented, it will have the opportunity of requesting a correction at the next meeting, or at least of recording at that time its concern that its or another party's comments were misconstrued or misrepresented in some way.

1. See 'Without Prejudice' section on p.247.

Mitigation

Mitigation in a construction dispute context can best be described simply as the effort put into the reduction of a risk by minimising its chances of occurring or, if its occurrence is inevitable, then by minimising its likely effects. Mitigation is 'avoiding the consequences of a wrong whether in tort or breach of contract'.[1]

In terms of the contractor's role in mitigation, its responsibilities and duties are to seek mitigation of its potential losses (time and money) and to ensure that it does nothing that would increase those losses. If the contractor fails in its duty to mitigate its loss then it cannot seek recovery of that loss. That is to say that if it fails to take reasonable mitigating action (or takes action which is inconsistent with or opposes the principles of mitigation) it cannot claim recovery of the losses that would have been avoided had it taken such actions (or had it not taken such inconsistent or opposing action).

The contractor's mitigation obligations in respect of time considerations are generally to take reasonable measures to reduce the potential delay. This could be by, for instance, re-sequencing programmed work events. If it merely sat back and let the contract run into delay without trying to take some measures to mitigate, unless of course no such measures were available, then it could not seek payment as compensation for the contract prolongation. If, however, as a result of re-sequencing to mitigate delay it incurs additional costs, then it would be entitled to seek to recover those costs. This would hold true even if the mitigation attempts proved to be abortive or if the end result was a claim greater in value than it would have been if the contractor had not attempted mitigation.

Unless the contract contains express wording to the contrary (which could include a direction that the contractor is to use its 'best endeavours' to prevent delay), the contractor, in its mitigation of an employer's delay, is not required to add extra resources or to work outside its planned working hours, although it may so do if the employer agrees to pay for the additional costs of such extra measures. The addition of the words 'best endeavours' in any contract clause requiring the contractor to mitigate the effects of delays could be interpreted as such express wording to the contrary, although it is held that 'reasonable endeavours' would not be so interpreted.

As regards loss and expense claims, the principle that the courts in England and Wales apply when assessing damages is that where a party suffers loss as the result of a delay for which it was not responsible it is entitled to be placed in the same situation, as far as is possible, that it would have been in financially if the delay had not occurred. This is the generally applied Common law rule with respect to breach of contract events.[2]

This rule applies to the assessment of damages rather than the conduct of the parties. Despite the general impression to the contrary, it has been held that there is no legal duty which requires the contractor to mitigate its losses. However, where it did not take mitigating action it would be unable to pursue a claim for losses incurred:

> *A plaintiff is under no duty to mitigate his loss, despite the habitual use by the lawyers of the phrase 'duty to mitigate'. He is completely free to act as he judges to be in his best interest. On the other hand a defendant is not liable for all losses suffered by the plaintiff in consequence of his so acting. A defendant is only liable for such part of the plaintiff's loss as is properly to be regarded as caused by the defendant's breach of duty.*[3]

If there are express clauses in the contract requiring the contractor to mitigate the effects should such an event occur then this would impose a contractual obligation upon the contractor and failure to take mitigating action would amount to a breach of contract.

To the extent that a contractor's mitigation attempts are successful, it is left in a position where it can still claim those losses which it has been unable to mitigate, should there be any, i.e. losses over and above those mitigated.

Let us consider the following example:

- Step 1: The contractor is working to programme, getting ready to erect steelwork to be supplied by the employer.
- Step 2: The employer informs the contractor of an anticipated one-month delay in arrival of the employer's supply steel at site.
- Step 3: The contractor already has a job-specific crane on site ready to start erecting the steelwork.
- Step 4: The contractor's mitigation would be either to send the crane back to the hiring company until needed or, if it was its own plant, to relocate it to another contract (if the contractor had need of one of that particular type and size of crane at that time).
- Step 5: The costs of demobilisation and remobilisation would be claimable losses; any sub-contractor losses resulting from this situation would also be recoverable.

If it is difficult to anticipate the likely extent of the delay period at the outset of the delay then the contractor has to make a commercial decision as to whether to take mitigating action or not. If it decides positively (i.e. to take mitigating action) only to find that the period of delay is shorter than anticipated and the plant has to be returned so much sooner that the total cost of demobilisation and remobilisation is more than the cost of standing time would have been had the plant not been demobilised, the contractor would still be able to recover the full extent of its losses.

The contractor's claim in such circumstances would be for the cost of the resources awaiting demobilisation (where it was feasible and not unreasonable for the contractor to be cautious about moving plant off site until it was clear what the likely delay might be), together with the costs of demobilisation, remobilisation and any other sub-contractor or other trade costs incurred as a result. The contractor cannot claim for the resources during the period when they were off site or elsewhere deployed.

1. Harvey McGregor QC, *McGregor on Damages*.
2. *Robinson* v *Harman* (1848) 1 Exch. 850.
3. *British Westinghouse Electric & Manufacturing Co. Ltd* v *Underground Electric Railways of London Ltd* (1912) AC 673.

Monthly Reports

It is usual for the employer to require the contractor to submit monthly reports relating the progress of the works, recording the resources utilised during the previous month, any problems encountered and overcome, programme variances both historical and forecast, commercial and quality issues and so on.[1] This is an essential tool for the employer, enabling it to plan its cash flow with greater accuracy, to plan whatever resources it is contributing to the project and to plan its taking over of the works. It is also an essential element of support documentation, which can be utilised by either party in either a claim or claim defence situation. Contemporaneous records prepared and submitted officially by one party to another at the end of each month are taken much more seriously by those involved in settling disputes than documents prepared after the event.

It is suggested that the following list, though not exhaustive, represents the minimum range of topics that should be included in the monthly report of a major project. No one person will prepare each and every topic on its own and so we should take responsibility for coordinating with the responsible engineers, planners, administrators, QA/QC personnel etc. to ensure that the report is complete, factual, accurate and covers all the topics which we consider to be pertinent not only to the employer's needs but also to the perceived needs of the contractor bearing in mind the possibilities of any known or anticipated dispute issues arising either at the time or at a later date. The range of topics suggested is:

- covering letter and executive summary;
- summary of engineering progress;
- summary of procurement progress;
- summary of expediting, inspection and fabrication progress;
- summary of construction and erection progress;
- summary of commissioning and tests on completion progress;
- an update of all safety statistics including a statement of the number of accidents or other incidents during the reporting period resulting in damage to equipment or plant and injury or death of personnel engaged in work on site, or incidents that represent a near miss to these events, together with details of such accidents and incidents, and statements of corrective actions taken;
- document index marked up to show document status;
- a statement of the number of site personnel engaged in the work during the reporting period (at site, at the lay down areas and in the project office) including an update of manning histograms and, where relevant, a statement of construction equipment at site during the reporting period;
- a list of late activities compared to the respective originally planned (early finish) dates;
- explanations for late activities and other problem areas which are having, or are likely to have, an impact upon the project schedule/critical path network;
- details of the measures proposed to resolve the problem areas and bring the

overall project back on schedule or further improve on current progress;

- outstanding interface data and measures proposed to expedite the issue of critical interface data;
- progress report on the milestone achievement schedule with date forecasts in respect of the milestones which are expected to be reached within the following three months;
- purchasing schedule associated with major or significant items of equipment, marked up to show status of manufacturing and procurement activities and status of shipments;
- shipping documentation, including bills of lading, commercial invoices, packing lists etc. and noting ETD[2] dates, ETA[3] dates, port of expected entry etc.;
- record of customs, import taxes and other duties paid during reporting period;
- documents (including drawings, material submissions, method statements, test procedures etc.) issued during the reporting period with status (i.e. acknowledged, accepted, rejected, under review, no response);
- list of permits, approvals, licences etc. applied for from all local, state or national government authorities or public service undertakings during reporting period, including status (i.e. acknowledged, issued, rejected, under review, no response);
- copies of inspection reports issued during the reporting period highlighting those which identify any deviations from the quality standards set out in the contract and including a statement of corrective actions;
- copies of quality assurance audit reports issued during the reporting period highlighting those which identify the need for corrective actions together with evidence of the implementation of corrective actions;
- colour photographs showing the progress of the works, an agreed number of which should be taken from the same agreed spot each month of significant parts of the works to provide an easily understood pictorial representation of progress;
- update of progress S-curves;[4]
- correspondence schedule listing all correspondence issued during the reporting period between the employer and the contractor, indicating actions required;
- record of all change orders requested and issued during the reporting period with cost and time effects;
- record of all RFI[5] submissions during the reporting period with status report (i.e. acknowledged, answered, under review, no response);
- record of all claims submitted to or by the employer during the reporting period, indicating their status (i.e. acknowledged, accepted, rejected, under review, no response) and their values;
- record of all official notices issued in accordance with the contract requirements;
- a statement of progress made in the preparation of all documentation to be handed to the employer at milestone dates or at other dates specified in the

contract (such as at Provisional Taking Over, Final Acceptance etc.);

- OS & D[6] reports issued during the reporting period;
- copy of the project schedule/critical path network marked up to indicate the progress achieved to date;
- financial status, detailing payments claimed, payments approved, payments received, payments outstanding and payments in dispute.

1. A typical 'Monthly Progress Report' format (used on a major engineering project) is given in Appendix 7 on p.295 – this can be adapted to suit different types of project and particular contract requirements as appropriate.
2. Expected/Estimated Time of Departure.
3. Expected/Estimated Time of Arrival.
4. 'S' curve – a display of cumulative costs, labour hours or other quantities plotted against time.
5. Requests for Information – see separate section in this book on p.203.
6. 'OS & D' reports are Over, Short and Damage Reports – prepared during inspection of materials and equipment delivered to site. These are either used to claim replacement and/or repairs by the relevant suppliers, or to claim against contractors involved in delivery and/or receiving, or to support insurance claims – see OS & D Report form in Appendix 6 on p.293

Negotiations Procedures

We need to be well versed in negotiation skills since negotiation plays a significant part in our job. This does not only involve the negotiation of the tender from the date of enquiry to the signing of the contract agreement, although this is obviously a significant negotiation exercise which needs our closest attention, but also includes the day to day negotiations as the contract works proceed on such matters as the extent and cost of change orders, extensions of time, disruptions, delays, sub-contractor and supplier issues, permits, licences, taxes, duties and other levies, insurance claims and so on.

There are basically four stages to most negotiations that we should recognise and become familiar with. These can best be defined as:

- preliminary introduction of parties to negotiation;
- agreeing objectives of negotiation;
- negotiation process proper;
- review and finalisation of agreements reached.

These stages of negotiation can be expanded upon thus:

- In the preliminary introduction stage the parties to the negotiation get to know each other. The team members establish the strengths, positions and levels of authority of their opposite numbers and are better able to plan their own strategies as a result.
- After the initial introductory phase it is important to agree on the objectives of the negotiation, not only as regards one's own team but also as regards the opposition. This is the time when an experienced negotiator should be able to discover where the priorities of the opposition lie and as a result be in a position to modify its own team's approach to maximise their own chances of success.
- When the negotiation process begins in earnest it should be split into two distinct stages. The first stage will be to resolve the technical issues. The full negotiation team should sit in on this part of the discussion so that the senior negotiators are aware of the issues in dispute and the attitudes of each side's technical experts in this respect. In the next stage the team may be reduced so that only those who have the authority to settle are involved. The technical issues should have been resolved in the previous stage, leaving the commercial time/cost issues to be resolved in this last stage of negotiations. It is useful to have members on the negotiating team who are well versed in costing and scheduling in order that the impact of the concessions and agreements made can be instantly evaluated.
- It is important to summarise the agreements reached as soon as possible after the negotiations have been concluded and it is wise to have a draft summation prepared while the parties are still present in order that it can be reviewed by the parties to the negotiation while the agreements reached are fresh in their minds. If this is left until the negotiators have returned to their bases and discussed the negotiations with their office management we might find that agreements

become clouded by comments made by our colleagues and by management not directly involved in and unfamiliar with the negotiations. The draft summation should be signed as a true record of the agreements reached before the negotiators leave the negotiation arena. The formal document confirming the changes agreed upon can be prepared subsequently but it is wise to agree at the end of the negotiation proceedings upon a timescale for the preparation and signature of this final document.

Negotiation is as much about tactics as it is about being convinced of the strength of one's own case and being totally familiar with all the commercial and technical aspects of the subject matter of the negotiations. The first stage of the negotiation process mentioned above will help define the tactics to be employed. It is possible to categorise some of the tactical approaches most common in the industry:

1. Simple negotiation on facts

The most basic negotiation approach is to concentrate on the facts of the negotiation and to succeed in obtaining the best results possible. If the negotiations are cost related this approach hinges upon the negotiator knowing the pricing structure in detail and having the capability to instantly evaluate proposed changes. There has to be an element of bluff and counter bluff on both sides but generally the facts resolve the issue.

A second feature of this approach is the ability to compare similar prices in respect of similar situations. In order to achieve success at this the negotiator must be familiar with price levels in the area where the work is being carried out and with labour rates, working conditions and availability.

2. Pressure

It is sometimes necessary to put the other party under some degree of strategic pressure in order to speed up the agreement process. This can be in several forms. It could be that the negotiation is in relation to the first of many similar orders or projects that the parties might execute together. In such a case it is easy to use the 'carrot' of further work, which the other party would like to be awarded, to obtain concessions.

Alternatively it could be that there are deadlines that have to be met, either because the buyer is under pressure to complete the works and so needs to purchase a particular part quickly or because the seller has surplus stock that is getting close to its 'sell by' date. If there are no deadlines that can be used to pressure the other party then the experienced negotiator should be able to create its own pressures and to press the other party into conceding points that it might not concede in a less stressful negotiating environment.

3. Seeds of confusion

This can include creating an element of surprise whereby changes are introduced without warning and the other party does not have sufficient time to fully appreciate or discuss the full consequences or merits of them with its team before proceeding with the negotiations. The new offer might appear reasonable on the face of it but subsequent ramifications might be felt in other areas not considered by the other party. It has also been known for one party to deliberately confuse the other by using distorted figures or information, or by raising an issue and then dropping it again, in order to sow the seeds of confusion. Use of this tactic has to be carefully balanced with considerations of ethics, as no negotiator should be deliberately unethical in its approach.

4. Stalling tactics

These might be useful to give either party time to go back to its negotiation team and their superiors in order to evaluate the present position. These can be introduced by simply stating that higher authority is needed before agreeing to the proposals currently on the table or by a simple request for a short recess in order to review the current position.

5. Togetherness

If the negotiations are being carried on in a friendly and cooperative manner there is scope to plead to the other party's sense of fair play and reasonableness and to emphasise the common goal of a mutually acceptable settlement. Both parties need to maintain their spirit of reasonable cooperation if they are to succeed in taking their project through to completion, and it can be put forward that reaching an amicable settlement here is essential to the well being of the project. The 'togetherness' vote comes into play at its best where the party promoting it has conceded several points already. It is relatively easy in such circumstances to make the opposing party feel that it is being unreasonable in not conceding one point when it is patently aware that the other party has already conceded so many, even if those points conceded are relatively minor ones. When using this straightforward 'give and take' approach, the successful negotiator is the one who gives less than it takes.

6. Sympathy

I once worked with a 'loosely formed' consortium and was employed by one of the members to manage the various aspects of the business. Consortium meetings were complicated affairs with pre-meetings between the strongest members to decide strategies that would then be put to the full consortium meeting, and subsequently approved.

The pre-meetings were preceded by pre-pre-meetings between my employer and his team. This would invariably result in one of the team being told that they were going to be hauled over the coals that day but not to take any notice of it. This was all part of his particular negotiating ploy. He would make himself appear to disagree with the progressive ideas of his team and to be holding back the development of the consortium's business by his inflexible and unreasonable attitude, which also appeared both heartless and ruthless. Needless to say the consortium members would jump on him, support the team's suggestions (which in fact were at least in part his own) partly out of sympathy for the team and partly for the perceived good of the consortium, which is just what he wanted.

I have subsequently seen US companies play the sympathy card by employing female Project Managers/Commercial Managers to put forward contentious issues to their Japanese 'partners' because they know that Japanese sensitivity will not allow them to confront a woman in the same way as they would a man. All is not what it appears, however, and that is one ploy that has worn a bit thin. The Japanese now have female managers who are a match for anyone. However the 'sympathy' card is played, we must make absolutely sure that doing so does not offend the law of the land or the social obligations of that land and that it does not cross the bounds of professional ethics in any way.

7. Last resort

The last resort is to threaten either litigation or arbitration. This threat is usually made by the underdog as in both litigation and arbitration there is a tendency to support the party with less weight behind it. In many cases the very threat of litigation or arbitration will in itself trigger a resolve by the other party to conclude negotiations without taking up either litigation or arbitration.

The one tactic to avoid at all costs is anger. It might work with a certain limited success in respect of some minor issues but not with respect to major issues. Apart from outpourings of anger tending to obscure reasonable thought as regards the issue in question this tactic also sours future relationships. When negotiating with business colleagues from the Far East or South-East Asia this is particularly important as to resort to anger against them is to lose both face and stature. The negotiator who resorts to anger has lost the argument and has lost all future respect as a negotiator.

Client negotiators from the Middle East tend to use time as their negotiating ploy. They are generally in no rush (I have heard the Arabic phrase 'Insh'Allah' referred to as 'maniana without the urgency') and like to keep the other party waiting, during which time they hope that initially it will become frustrated with waiting, and then that doubt will set in as to the strength of its position. After a while they hope it will get to the point where the other party will be prepared to settle below its original intentions just to agree something and be able to get on with the project.

The Japanese have traditionally avoided confrontation in respect of business disputes and their primary negotiating ploys revolve around misunderstanding and denial.

They are historically unfamiliar with formal contracts since Japanese domestic business developed using informal acknowledgements of contract agreements, and their initial reaction to an issue might well take the form of denial that they meant a particular term of the contract or specification to apply, either through misunderstanding or unfamiliarity with the term employed.

This attitude is similar to the Middle Eastern one whereby the signing of a contract is considered to represent the start of negotiations. The Japanese also tend not to disclose mistakes, since in their culture to do so would be accompanied by a loss of face, and they will prefer first to deny an issue exists and then to make excuses which either cast the blame elsewhere or suggest that the 'alternative method/material' that they have employed is an improvement on the original intention.

Employers negotiating under such circumstances need to be firm in demanding full disclosure of all the facts relating to any anticipated delays, specification or QA/QC issues and in implementing the various powers and remedies available to them under the contract. If in doubt the secret is to check and double check any suspect issue from as many different angles as possible until satisfied that the full facts have been disclosed. In saying that, however, it is extremely unlikely that a Japanese company would not honour its contract at the end of the day even if this involved additional expenditure on its part. Unfortunately the same reliance cannot be placed upon all Middle Eastern employers.

Notices

Conditions of contract invariably require the contractor to issue notices at a specific point in time or within a specified period of time in respect of certain situations that it might find itself in, either by chance or in the course of its execution of the contract. We must be vigilant in ensuring that all notice requirements are strictly adhered to.

Certain notices have to be served by the contractor in order for it to avail itself of remedies as specified within the contract (such as extensions of time, change orders and so on), others are pre-requisites of formalities which the contractor has to go through upon the occurrence or completion of certain specified events (for example, achievement of milestones for payment purposes, certificates of completion etc.). The employer also has responsibilities as regards issuing notice, the most important of which is probably the notice to proceed with the contract works. This section deals principally with notices in respect of extension of time claims.

Where the contract states that the issuing of a notice at or within a specific time is a condition precedent then a total failure to issue a notice in the proper manner and at the proper time will be fatal to the contractor's claim. Each case has to be judged on its merits and depends to a considerable extent on the precise wording of the contract in respect of notices and claims but in general, where it is not specifically noted as a condition precedent to a claim, its late submission should not result in the contractor losing its rights to claim either under the contract, or possibly as breach damages assuming that the contractual rights are not specifically noted as exclusive remedies.

Clauses which purport to make the issuance of a notice within a certain time frame a prerequisite for a claim of time at large are worthless because once time is at large the notice provisions no longer have any contractual relevance and are in effect defunct. There is an interesting discussion later in this book[1] concerning notices which are conditions precedent to the granting of extensions of time which in turn are due to acts of prevention by the employer, the discussion relating to the case where such notices are not given. It is put forward that by applying the principle of prevention, the extension of time clause would be inoperable and time would be at large.

Where no specific time frame is given for the serving of a notice but rather words such as 'as soon as possible' or 'within a reasonable time' are used, the situation is not so clear. It has been held that an insurance claim for an incident that occurred on 12 June 1989 but which was not submitted until 4. July 1989 was out of time because the relevant clause in the insurance policy required claims to be made 'as soon as possible', which three weeks later was obviously not considered to be, under the particular circumstances.[2] Basically, the interpretations of 'as soon as possible' and 'within a reasonable time' depend on the exact wording of the contract and the circumstances surrounding the particular event or occurrence in relation to the contract, its location, the parties concerned and so on.

Frequently contractors fail to give proper notices and subsequently rely on reports given at site meetings to support their claims on the basis that the 'notices' given and recorded at the site meetings were given in due time. It has been held that, unless the parties agree at the outset of the contract that the obligation to submit formal written notices shall be waived in favour of delays being recorded in site meeting minutes, recording delays in site meeting minute records will not constitute 'good notice'.[3] We must ensure that both the employer's and the contractor's project teams are aware that this approach is not permitted and must ensure that the proper notices are served in accordance with the contract requirements.

Notices should also adhere rigidly to the requirements of the contract in respect of their exact wording and the facts which are to be included in them. Even where notices are given on time, if a specific fact, such as a date when the subject matter of the notice is to come into effect, is omitted then the courts will invariably find that such omission renders the notice defective.[4]

In most cases it is safer to stick rigidly to the notice requirements of the contract, whether they are in respect of claims for change orders, extensions of time, interim payments, insurance claims etc. or whether they are in respect of material and equipment quality control tests, performance tests and so on. If the rules are strictly adhered to, any claim emanating from that particular event or occurrence will be all the stronger as a result.

We should, at the outset of the contract, prepare a schedule of all the notice requirements in the contract and we should instruct all members of the project team, including site managers, section and discipline leaders, quality control and safety personnel etc. in those notices which are of particular concern to them. During the course of the contract we should be ever vigilant in monitoring the issuance of notices at the appropriate times. If we fail to do so, our job at contract close out stage or whenever we are involved in resolving outstanding claims will be far more onerous and will result in a less than successful conclusion to the commercial aspects of the contract.

A particular contract I was involved in for the construction of a large energy-related project in Australia included a clause that started off:

> *It is a condition precedent to the Contractor's entitlement to make a Claim that the Contractor has fully complied with each and every requirement of this clause.*

The clause continued with a number of sub-clauses prescribing events and situations that the contractor was obliged to issue notices in respect of and setting out periods within which such notices should be issued. It then concluded by stating:

> *If the Contractor fails to comply with each and every requirement of this Clause in relation to any particular fact, matter or thing (as the case may be), that may constitute a Claim, then the Contractor will be absolutely barred from making and*

will be deemed to have irrevocably waived any right to make (and the Owner will have no liability whatsoever in relation to) any Claim (to the maximum extent permitted by law) against the Owner arising out of, or in any way connected with, the particular fact, matter or thing (as the case may be). This Clause does not, of itself, create an entitlement of the Contractor to make a Claim.

Contractors, when faced with such a clause, should work with two basic concepts in mind, namely that:

- too much is better than too little; and
- sooner is better than later.

In saying this I mean that they should serve notices well within the limits set by the contract, not at the last minute, and should keep all possible records. If they are in doubt as to whether a notice should be served, they should serve it anyway; if they are in doubt as to what records the employer might require, they should obtain a ruling from it at the earliest opportunity. If the employer's ruling is unsatisfactory, incomplete or unclear in defining such records then the contractor should keep more records rather than less.

1. See 'Time at Large' section on p.238.
2. *Kier Construction* v *Royal Insurance (UK) Ltd* (1992).
3. *John L. Haley Ltd* v *Dumfries and Galloway Regional Council* (1988).
4. *Embankment Place Hotels (Blackfriars) Limited* v *(1) Blackfriars Hotels Limited; (2) Six Continents Hotels (UK) Limited (formerly Holiday Inns (UK) Limited)* (2002).

Omission of Work – Loss of Contractor's Profit

The question is often raised when preparing or defending claims on international contracts as to whether or not the contractor is entitled to claim for loss of profit on omitted work. There is no clear-cut answer and the merits of each individual case depend upon the purpose and circumstances of the omission, the applicable law and the wording of the actual contract clause under which the omission is being made.

If the contract makes allowance for the recovery of an element of profit in addition to other compensation, this may be calculated using the average profit earned by, and available in the audited accounts of, the contractor for, say, the three or five-year period immediately preceding the event which gave rise to the dispute and subsequent claim. Alternatively, if the parties agree, the profit percentage used in the tender build up may be applied to the omitted work.

One important principle applicable to construction contracts is that a contractor must be permitted to carry out and complete the whole of the works it has contracted for, subject to reasonable variations being agreed between the parties. Under normal circumstances, and in the absence of express words to the contrary, the employer cannot omit work from the contract specifically to give it to another contractor. If it did so it would be in breach of contract and the contractor would be entitled to recover its loss of profit on the omitted work as part of its action for breach.

Any omission of work by the employer's representative must be made in accordance with contract clauses empowering it to do so. If there were no such clauses than the employer would again be in breach of contract and the contractor could claim loss of profit as part of its action for breach. A typical clause that would go some way to safeguarding the contractor's interests, as regards omission of work by the employer, would ensure that:

- the contractor was not bound to carry out a change order involving an omission of part or parts of the work where the aggregate effect of compliance would be to decrease the contract price by more than (say) 20 per cent;
- the employer and contractor would meet to discuss an equitable settlement on the proposed reduction in work scope, sufficient to:
  - compensate the contractor for losses, expenses and costs incurred or to be incurred as a result of the omissions; and
  - enable the contractor to implement the omission instruction.

Where contracts include the words 'direct loss and/or damage' in the compensation clause (unless there is anything elsewhere in the contract to say otherwise) these words can be interpreted as covering the contractor's loss of profit.[1] It might also be possible to utilise the rate adjustment provisions of the contract, if there are any, to argue that the omission of portions of work have made specific rates or prices unreasonable or inapplicable in that their application no longer enables the same level of profit return contribution to the contractor's overall company turnover which it was anticipated, or which it could reasonably have been anticipated at tender stage, that such works would contribute. This argument has to be considered strategically bearing in mind the profit earning capacities of the work omitted and the profitability of work added in the form of change orders or variation orders.

In Civil law jurisdictions where 'rules' are relied upon to a far greater extent than in Common law jurisdictions, the situation is more clear-cut as regards loss of profit claims. Under the French Civil Code, for instance:

> *Damages due to a creditor are, as a rule, for the loss which he has suffered and the profit which he has been deprived of . . .*[2]

and

> *Even in the case where the non-performance of the agreement is due to the debtor's intentional breach, damages may include, with respect to the loss suffered by the creditor and the profit which he has been deprived of, only what is an immediate and direct consequence of the non-performance of the agreement.*[3]

Thus where the omission was of such an extent that the contractor had a legitimate claim against the employer, under the French Civil Code jurisdiction the contractor would clearly be able to recover loss of profit as a part of his overall damages claim.

Under the German Civil Code, a significant omission of work that was previously included in the contract between the employer and the contractor would give the contractor the right to claim 'adaptation of the contract'.[4] This would be its opportunity claim loss of its anticipated profit on the basis that the circumstances upon which the contract was based have materially changed after conclusion of the contract and the parties would not have concluded the contract or would have done so upon different terms if they had foreseen that change (i.e. the reduced scope of work). Also under the German Civil Code,[5] if for some reason it is not possible to adapt the contract, or if adaptation cannot be reasonably imposed upon a party to the contract, the disadvantaged party may terminate the contract on notice.

1. *Wraight* v *P. H. & T. Holdings Ltd* (1968) (unreported).
2. French Civil Code, Article 1149.
3. *ibid.*, Article 1151.
4. German Civil Code, Article 313 – Interference with the basis of the contract; part (1).
5. *ibid.*, Article 313 – Interference with the basis of the contract; part (3).

Pre-Tender Information Review

It is extremely important that the project team investigate the site and its location thoroughly during the bid process. This should be carried out along predetermined lines of investigation and it is suggested that a check sheet similar to that set out in Appendix 3[1] be used for each project. The items noted on the review sheet are by no means exhaustive and this list can be adapted to suit each individual project and site location as necessary. The report should be prepared as a result of the contractor's project team spending time in the area where the project is to be executed and researching all aspects of the prospective works, possibly in cooperation with the employer or its local representatives.

This is a most useful tool in the future management of the project as, apart from the help it provides to those involved in the day to day management of labour, material, plant and equipment resources and logistics, it helps those of us involved in the commercial and legal aspects of the project to identify changes in circumstances between the initial bid investigations and commencement and execution of the project, which in many cases could be several years apart.

Particular attention should be given when carrying out the research to circumstances affecting the supply of local labour, materials, plant and equipment, including wage agreements, conditions of employment (including usual working hours, overtime and bonus norms), quality and availability of both locally manufactured and imported materials, and the local situation regarding availability of plant and equipment, including the availability of service and spare parts. The research team should obtain local information regarding sub-contractors in the various disciplines that will be needed on the project works, covering the quality of work performed by them, productivity standards, availability, reliability, capacity and so on.

The importance of satisfaction regarding the available site survey information will be seen from the later section in this book on sub-soil investigations[2] and the contractor will have to make a reasoned judgement at the bid stage whether to rely on the sub-soil information provided by the employer or whether to engage its own specialists to carry out a detailed survey.

Where the contractor will bear the sole responsibility for the sub-structure works, drainage etc., and where the contract specifically places this responsibility upon the contractor, the contractor would be wise to commission a detailed site survey to establish the nature of the sub-soil, the site levels, the ground bearing capacity, whether piles will be needed, whether the ground is of natural formation of whether it is comprised in whole or in part of fill material, whether de-watering techniques will be required, etc. in order to minimise its risk exposure.

Another important aspect of the review is in regard to port unloading facilities and capacities, customs clearance procedures, local transportation and logistics in general. The time taken for imported materials, plant and equipment to be unloaded, clear customs inspections or other formalities and reach site after arriving at the port of discharge can be critical if the programme dates are to be met. It is wise to discover all the likely risks before the programme is finalised in order that contingency float can be included in the programme to cover whatever eventualities are found to occur in that particular area with any degree of frequency.

It might be that the port is working over capacity and that there is frequently a backlog of shipments waiting to be unloaded. It might be that employer insists on an independent inspection of shipped goods before they are unloaded, in which case it would be necessary to find out where their inspector was based, the inspector's average workload and availability, costs of demurrage and so on. If the inspector covers a wide area then there might be a need to arrange an early warning system whereby adequate notice of the shipment's arrival could be forwarded to the inspector, reducing the risk of delays in this respect, or to include contingency sums in the bid to cover the possibility of demurrage charges being incurred while awaiting inspections.

It is also particularly important when dealing with international contracts in countries with which we are not familiar to obtain local knowledge of the religious and business customs of the area. On a project I worked on in Thailand we ignored a request to make a 'donation' to a local body and lived to regret our decision. We were, some days after rejecting the invitation to donate, celebrating the successful inauguration of the first phase of our project when there was an almighty explosion followed by a fire in a large transformer, rendering it useless and delaying the second phase of our project by more than six months. For the next three years we 'donated' generously to all local bodies, had each significant milestone event blessed by representatives of the local religious faith and had no further 'accidents'.

We had no proof of the link between our refusal and the incident but neither did we have any reason to doubt our conclusions. If our company had carried out a thorough pre-tender investigation, and if they had passed that on to the team managing the project, we would have been aware of the customs in the area, particularly those affecting foreign companies operating on their soil. We would have followed the precedents established by our forerunners and this incident might have been avoided.

In some countries, particularly in the Middle East, governments often require that certain percentages of the contract works and material purchases be awarded to local companies. In addition to stating overall percentage requirements, the work packages or materials affected by these rules are often specified by the government, but the local manufacturers concerned cannot always provide materials or workmanship equivalent to the standards specified in the contract and are not always competitively priced. This will have to be investigated and solutions found prior to finalising the bid.

It might be possible to convince the authorities that the local manufacturers or sub-contractors are not able to comply with the specifications, necessitating the importation of resources, failing which the bid will have to be adjusted to compensate for any additional costs of using local resources and, possibly, for the provision of additional quality control measures to overcome the local deficiencies.

The form set out in Appendix 3 provides a comprehensive guide to the extent of information it is advisable to obtain on major international projects. It cannot be over-emphasised that it is far better to obtain too much information than too little. The bid preparation team require all the help they can get in order to prepare a realistic bid. It is left to their judgement to decide whether a particular item of information requires a cost to be included in the bid, whether the risk can be borne by the company or whether the risk can be passed on or covered by insurance.[3] Without such a comprehensive pre-tender site review, reasoned judgements would not even be possible.

If the tender bid is successful the project execution team should be given a copy of the Pre-Tender Site Review together with a report from the bid preparation team stating how the various findings of the review were dealt with in the bid. The project execution team should refer to these documents frequently as each new step in the contract is reached.

1. See Appendix 3 – 'Contractor's Pre-Tender Site Review' on p.279.
2. See 'Sub-soil Investigations Included with Tender Enquiry' section on p.233.
3. See 'Risk Allocation, Assignment and Assumption' section on p.217.

Prevention

The principle of prevention basically stands to prevent one party claiming that the other party is in breach of a contractual obligation where the first party has in fact impeded the second party and thus prevented it from performing that obligation.[1]

This is particularly relevant when considering access to the site in order to commence the execution of the contract works and to maintain continued access during the construction of the works. Should the employer deny the contractor access at any given time such that it could not commence or continue with the contract work, this would be classed as an act of prevention and would usually be covered by a clause in the contract entitling the contractor to an extension of time in respect of a delay caused by the employer.

If no such extension of time clause is included in the contract or if the extension of time clause is not properly operated then the time for completion set out in the contract no longer applies and the contractor is obligated to complete within a reasonable time. The employer is in that case no longer entitled to apply liquidated damages on the contractor even if the contractor is responsible for part of the delay.[2] It is virtually impossible to draft a contract that effectively holds a contractor to a time for completion regardless of any acts of prevention or delay that the employer may cause the contractor.

There is implied in every contract an agreement that neither party shall do anything that could hinder or prevent the other party from fulfilling its contractual obligations.[3] Accordingly the innocent party is given the right in such circumstances either to sue for damages if a breach of a warranty or to determine the contract if a breach of a condition of the contract.

The act of prevention need not, however, be a breach but could be a legitimate action under the contract, such as ordering additional work. Where this in turn prevents the contractor from completing within the stipulated time and where there is no contract provision for extending the contract period appropriately, or if the extension of time clause is not properly operated, such an action could be considered as an act of prevention and the party ordering the extra work could not claim any penalties or liquidated damages for delay in completion beyond the stipulated time for completion.[4]

Some contracts make provision for the payment of a bonus to the contractor, conditional upon its early completion of the works. If the employer in such a contract then issues variation orders for additional work which, because of their nature, inevitably extend the time for completion beyond the agreed early completion date upon which the bonus would have become payable, the employer could be held in breach of an implied non-hindrance term in the bonus agreement and could be liable for all or part of the bonus which the contractor thereby failed to earn.[5]

1. *Amalgamated Building Contractors Ltd* v *Waltham Holy Cross UDC* (1952) 2 All ER 452 CA.
2. *Peak Construction (Liverpool) Ltd* v *McKinney Foundations Ltd* (1. July 1970) 69 LGR 1.
3. *Banque Quilpue Ltd* v *Bryant* (1904).
4. *Trollope & Colls Ltd* v *North West Metropolitan Regional Hospital Board* (1973) 2 All ER 260.
5. *John Barker Construction Ltd* v *London Portland Hotel Ltd* (1986).

Quantum Meruit

'Quantum meruit' is a Latin expression the literal meaning of which is 'as much as has been earned'. In construction contract situations this is interpreted as 'the amount a party deserves' or 'what the job is worth'. When a person employs another to do work without any agreement as to the amount or method of compensation, the law implies a promise from the employer to the workman that it will pay it for its services (as much as has been earned or what the job is worth).

The plaintiff may suggest in its declaration that the defendant promised to pay it as much as it reasonably deserved, and then state that its trouble was worth a sum of money, which the defendant has omitted to pay. This is called an assumpsit[1] (promise or undertaking) on a quantum meruit. When there is an express contract for a stipulated amount and method of compensation for services, the plaintiff cannot abandon the contract part of the way through and resort to an action for a quantum meruit on an 'implied assumpsit'.

The need for us to apply the principles of quantum meruit can arise in relation to construction contracts, even though this should not generally be the case since there are usually adequate contractual provisions which cover most aspects of payment for work done, whether they are included within the original scope of work, added by means of a change order or agreed on a daywork basis. Exceptions, however, do occur. For example, in one particular case[2] where the main contractor issued its sub-contractor with a variation order covering certain additional work, the arbitrator held that the work covered by the variation order did not form part of the main contract and since the variation did not therefore constitute a valid instruction under the main contract, the sub-contractor was entitled to payment on a quantum meruit basis.

Quantum meruit claims presuppose that no contract exists as far as the specific work item to be valued is concerned, and can arise where:

(i) there is no agreed fixed contract price for the work carried out, even though there might be a general contract in place;
(ii) there exists an express agreement to pay a reasonable sum;
(iii) the contractor carries out work outside the agreed scope of work of the contract, as was the case in the example quoted above;
(iv) work is carried out while contract negotiations are proceeding but, in the end, the negotiations do not result in a contract being forthcoming. Work carried out during this negotiation period should be paid for on a quantum meruit basis;[3]
(v) a contract is brought to a premature end by a repudiatory breach.[4]

Quantum meruit payments must be fair and reasonable and generally claims may be valued:

- by taking account of the contractor's costs, where they are deemed fair and reasonable; or

- by adopting general market rates;[5] or
- by determining the value to the recipient of the work carried out by the contractor; or
- by reference to charges that a competitor for the said work would have incurred had it been appointed to execute the said works (where such an argument is valid and relevant).

It has been held also that, where work which was carried out while contract negotiations were continuing, and where those negotiations failed (i.e. no contract came into existence) but the work continued on a quantum meruit basis as noted above in example (iv) of where quantum meruit cases may arise, the contractor could not claim payment in excess of the value of his tender that the negotiations failed to agree upon and which included the works thus carried out. (i.e. the contractor could not benefit by the failure of the negotiations).[6]

Examples (iii), (iv) and (v) of where a quantum meruit situation may arise noted above, (i.e. (iii) where the instruction to carry out certain work does not constitute a valid instruction under the contract; (iv) where no contract comes into existence; and (v) where the contract comes to an end as a result of a repudiatory breach) are often referred to as 'quasi-contractual'[7] situations and in such cases quantum meruit payment is to be based upon the principle of restitution (taking into account the point noted in the preceding paragraph regarding example (iv)).

The obligation will be to reimburse the contractor for the value of the advantage, if any, received by the employer as a result of the work done. In the *Costain Tarmac Consortium* v *Zanen* case noted above, the court held that the proper way to evaluate the quantum meruit claim in this case was to reimburse the sub-contractor the value of the advantage, if any, received by the main contractor as a result of the work done, whilst at the same time considering the relative bargaining positions of the parties at the outset of the job. In the case in question the quantum meruit would therefore take into account a proportion of the (considerable) profits generated by the Costain Tarmac Consortium, and would also include charges such as the likely cost of mobilisation and demobilisation which a competitor would have incurred had one been appointed, even though the sub-contractor in question was already on site and incurred no such additional costs.

The difficulties of determining whether and how quantum meruit should be applied when a contract exists were discussed in a Canadian case[8] where the Nova Scotia Court of Appeal quoted *Goldsmith on Canadian Building Contracts* (4th Edn, Carswell, 1988) at p.4-26:

> *In order to determine whether a change of so fundamental a nature has taken place it is necessary to look, on the one hand, at the contract in the light of the circumstances that existed at the time when it was entered into, and, on the other, at the events which have occurred and which are alleged to have brought about such a change*

> *of circumstances. It cannot, however, be emphasised too strongly that hardship, inconvenience or material loss, whether caused by unforeseen difficulties or by some other reason beyond the control of the parties, or the fact that the work has become more onerous than originally anticipated, are not sufficient to entitle the contractor to regard the contract as terminated. It is equally clear that as long as the contractor continues to work under the contract and recognises its existence, however much he may complain about the hardship which it is causing him, he cannot substantiate a claim on a quantum meruit basis. Before being entitled to do so he must clearly indicate that he regards the contract as being inapplicable; at the risk of being wrong and being liable for substantial damages for failure to complete.*
>
> *One of the main difficulties in practice is that in the case of many contracts involving protracted work it is often difficult to assess the effect of changes, either individually or cumulatively, until much of the work has been done, or until the contract has been practically completed. If the work has in fact been completed under the contract, it will be too late to assert a claim for payment otherwise than in strict accordance with the terms of the contract, whatever losses the contractor may have suffered. It is, therefore, of the utmost importance that a contractor who believes that a change of sufficient magnitude has occurred, should assert his rights at the earliest possible moment. If he is properly entitled to regard the contract as having become inapplicable by virtue of the far-reaching effect of the changed circumstances, he must take an unequivocal position that the contract has ceased to be applicable and is no longer in existence. If the owner or his representative does not accept the contractor's position, the contractor must make his choice of continuing under the contract or ceasing all further work.*

This clear and concise explanation indicates the need for the project team to monitor closely all changes to the work in order that they are in a position to submit a claim notice at the earliest possible opportunity should the need arise to take the position that the contract has ceased to be applicable due to the magnitude of the changes proposed.

1. *Oran's Dictionary of the Law* – '"He promised"; an old word meaning a promise to do or pay something. Certain types of lawsuits had this name. For example, "indebitatis assumpsit" was "he promised to pay the debt", but it was based not on an actual promise but on the fact that money was owed, whether or not there was an actual promise to pay'.
2. *Costain Civil Engineering Ltd and Tarmac Construction Ltd* v *Zanen Dredging and Contracting Co. Ltd* (1996) 85 BLR 85.
3. *Kitson Insulation Contractors Ltd* v *Balfour Beatty Buildings Ltd* (1990).
4. The test for repudiatory breach is essentially whether the offending party's conduct demonstrates that it no longer intends to be bound by the terms of the contract.
5. *Laserbore* v *Morrison Biggs Wall Ltd* (1992) – in which Judge Bosher QC considered that evaluation on a cost plus basis was wrong and that the appropriate approach was to adopt general market rates.
6. *Lachlani* v *Destination Canada (UK) Ltd* (1996).

7. 'Quasi-contracts' under English Common law are recognised similarly under the French Civil Code – see French Civil Code Articles 1370 to 1386 – 'Of Undertakings Formed without an Agreement'.
8. *Turf Masters* v *TAG Developments and Dartmouth* (1994) 135 NSR (2d) 105; Chapman J.A., Nova Scotia Court of Appeal.

R

Reasonable Skill and Care

The generally accepted standard by which 'reasonable skill and care' is judged is the standard of the ordinary man exercising and professing to have that skill.[1] It has been defined as the level of skill and care appropriate to the expertise that the party holds itself out to possess.[2] Where a party professes to have a greater than ordinary degree of skill, greater in fact than it actually has, the level of reasonable skill and care to be expected of it would be judged on the level that person professed to have, not the lesser level of an ordinary practitioner in that field.[3]

In an international construction context, reasonable skill and care would be that level of skill and care which an international contractor could be expected to exercise in the performance of its obligations and duties taking into account the usual standards prevalent in, and the stated standards required of, the particular industry, business or trade concerned.

The level of reasonable skill and care applicable to a designer would include a duty to amend its design if it appeared defective during construction,[4] although the designer would not generally have a duty of care to continue to warn of potential dangers after its involvement in the project had ended.

By adopting the principle of 'reasonable skill and care' the requirement becomes one of merely doing one's best without guaranteeing success, as opposed to 'fitness for purpose'[5] which imposes upon the contractor a strict obligation to succeed, failure to succeed resulting in the contractor being liable for the consequences of his failure.

Whereas under the Common law of England and Wales, the reasonable skill and care obligation is generally an implied term, under Civil Codes this would be expressed in terms of a rule. The French Civil Code has two somewhat contradictory rules that apply.[6] These rules state:

- Article 1137: 'An obligation to watch over the preservation of a thing, … compels the one who is responsible to give it all the care of a prudent administrator.' *(bon père de famille)'*
- Article 1147: 'A debtor shall be ordered to pay damages, if there is occasion, either by reason of the non-performance of the obligation, or by reason of delay in performing, whenever he does not prove that the non-performance comes from an external cause *(cause étrangère)* which may not be ascribed to him, although there is no bad faith on his part.'

These Articles relate to two distinct types of contract situation. In the first the debtor has an obligation to take the measures that a reasonable man (bon père de famille) would take to achieve the intent of the contract and is therefore bound to no more than the exercise of reasonable care, and this is covered by Article 1137. This has been termed an 'obligation de moyens'.[7] In the second contract situation covered by Article 1147 the debtor's obligation is not only to show due diligence but also to achieve the result which it has promised. This has been termed an 'obligation de résultat'.[8] To escape an obligation de résultat the debtor must show that its failure to produce the promised result is due to an external cause beyond its control, termed a 'cause étrangère'.

The difference between these two types of obligation, whilst not precise, is to be found in the degree of the burden of proof required. In the obligation de moyens the burden of proof lies on the creditor whereas in the obligation de résultat the creditor only has to show that the result has not been achieved and the burden of proof then lies on the debtor to show a cause étrangère.

Article 1137 makes liability depend upon proof of fault whereas Article 1147 makes exemption from liability depend upon cause étrangère. As examples in everyday life, consider a sports team manager – he does not promise success and so his would be an 'obligation de moyens'. Conversely an aeroplane journey has a definite starting point and a definite point of arrival and so the airline by which you are travelling would have an obligation de résultat, which could be denied by, say, diversion due to strong winds that would be classed as a 'cause étrangère', beyond the control of the airline.

Under the German Civil Code, Article 276:

(1) The obligor is liable for deliberate and negligent acts or omissions, unless the existence of a stricter or more lenient degree of liability is specified or to be inferred from the other subject matter of the obligation, in particular the assumption of a guarantee or the acquisition risk.
(2) A person acts negligently if he fails to observe the relevant accepted standards of care.
(3) The obligor cannot be relieved in advance of liability for deliberate acts or omissions.

Furthermore under German Civil Code, Article 277:

> *A person who is liable only for failure to display the standard of care that he displays in his own affairs is not relieved of liability for gross negligence.*

It can be seen that the situation under codified systems of jurisdiction, particularly the French Civil Code, is more precise than under Common law systems yet the basic principles are similar.

1. *Bolam* v *Friern Hospital Management Committee* (1957).
2. *Chaudry* v *Prabhakar* (1989); *Freeman* v *Marshall & Co.* (1966) 200 EG 777.
3. *Wimpey Construction (UK) Ltd* v *D. V. Poole* (1984).
4. *London Borough of Merton* v *Lowe* (1981) 18 BLR 130.
5. See 'Fitness for Purpose' section on p.119.
6. French Civil Code, Articles 1137 and 1147.
7. Sometimes also referred to as 'obligation de prudence et de diligence'.
8. Sometimes referred to as 'obligation déterminé'.

Request for Information (RFI Submissions)

There has been a change in the way in which the RFI process has been used in recent years in that it has become a vehicle for far more than just obtaining answers to issues which are unclear, ambiguous or in conflict with each other in the project documentation. The original concept of the RFI process was to have a quick and easy way of resolving day-to-day issues between the contractor and the employer and the employer's design team. As projects have become more complex the scope for errors, conflicts, omissions, ambiguities and so on has become greater and the need for RFI submissions has grown proportionately.

Contractors have expanded the use of RFI forms so that they are no longer merely used for day-to-day question and answer situations. They are now being used as a means of making submittals, responding to notices given under the contract in respect of non-conformances, transmission of documents, and generally for corresponding on a day-to-day basis in a relatively informal manner. In effect it could be said that they are being used as a soft approach to preparing the ground for future claims submissions.

The initial impression given to those brought in to mediate or arbitrate on a dispute is that where there is a need for a seemingly excessive amount of RFI submissions there must have been some serious lack of finality in the documents that the contractor was given at bid stage. This is a particular danger when the contract is a so-called 'lump-sum' contract where their initial assumption will be that the contract had originally been agreed on the basis of a fixed lump sum for a set scope of work and/or design. Where there are vast numbers of RFI submissions this is bound, at least initially before all the facts have been propounded, to influence the mediators or arbitrators in a dispute situation, either consciously or sub-consciously, to consider that the scope of work/design must have been far less than complete when the so-called 'lump sum' was agreed.

We have to be careful to take all measures necessary to protect the party for whom we work whilst maintaining an overall sense of fairness in the interpretation and application of the contract. We must only use, or allow the use of, RFI forms where there are facts which need to be established beyond all reasonable doubt as regards the employer's intentions to be adduced from the drawings, specifications and other documents. Where drawings have to be transmitted to the employer for review and/or approval these should be accompanied by a specially prepared submittal form, not an RFI form.

Where the contractor responds to a defects notice this should be on a specially produced reply form or by correspondence, not on an RFI form Day-to-day correspondence should be in standard letter or fax formats, not on an RFI form. The contractor must resist the temptation to use RFI submissions as a means of building up a case against the employer on the basis of volume of numbers alone. If, however, there are a significant number of genuine RFI submissions, sufficient to cast doubt over the fullness and completeness of the information provided to the contractor at both tender and contract stages, then these must be analysed and a case put before the employer.

Such a case founded on a factual analysis of genuine RFI submissions would find favour with an arbitrator to a greater extent than one founded on false claims and fiction. We must be constantly aware of this and must manage the RFI submission process to ensure that the use of RFI forms is strictly controlled and managed in as fair and reasonable a way as is possible taking into account the various needs and interests of each party.

The whole abuse situation could, and should, be prevented by the provision of definitions in the contract documents as to what constitutes or requires an RFI form submission, a document submittal, a memo, a particular notice, a fax or letter or any other form of transmittal, covering when and how such a document should be used. Employers sometimes provide definitions and requirements in their standard contract documentation and some have standard forms and procedures that are to be used on each particular project. This is by far the best situation to be in.

If the employer has not so provided, however, then it is up to either the employer or the contractor, as appropriate to the particular procedure concerned, to prepare the necessary forms and to obtain the other party's agreement to their use. Once agreed, they should be abided by to a fault. A typical RFI clause in a contract should:

- empower the contractor to submit an RFI to the employer in order to determine a clarification or interpretation of any portion of the contract documentation that requires it;
- refer to the use of a standard RFI format that should be included in an Appendix to the contract;
- require the contractor to state clearly the issue in question, the reason why a clarification or interpretation is required and the contractor's own interpretation in a concise manner;
- empower the employer to either accept and respond to the RFI or to reject the RFI as being the wrong medium for transmission of the request noted, both within a set number of working days;
- entitle the contractor to claim additional time and costs where the employer fails to respond within the time set for its response and where, as a result, the contractor suffers a delay to project completion and incurs additional costs;
- entitle the contractor to request a change order where the response constitutes a change to the contract as defined in the contract, provided that the contractor submits its request within a set number of working days of the employer's response, after which the contractor's right to claim at a later date will be lost.

Restitution

Restitution is the act of giving something back which under contract law usually means putting the plaintiff back in the position it would have been in had the wrongdoing not occurred. The three separate basic types of restitution claim can be classified as:

- the reversal of unjust enrichment;[1]
- the prevention of a wrongdoer from profiting from its wrongdoing; and
- the vindication of property rights with which the defendant has interfered.

In order for restitution to be appropriate there must have been unjust enrichment. This is the basic principle that underpins the law of restitution:[2]

> *It is clear that any civilised system of law is bound to provide remedies for cases of what has been called unjust enrichment or unjust benefit that is to prevent a man from retaining the money of or some benefit derived from another which it is against conscience that he should keep. Such remedies in English law are generically different from remedies in contract or tort, and are now recognised to fall within a third category of the common law which has been called quasi contract or restitution.*

The main restitution claim we will come across in international construction contracts is 'unjust enrichment' but 'benefiting from wrongdoing' may also sometimes apply. This is where the defendant received a benefit as a result of the commission of a wrong and although the plaintiff was the victim the benefit was not obtained directly from the plaintiff.

In order to claim unjust enrichment we have to ask the following questions:

(i) Has the defendant received an enrichment?
(ii) Was the enrichment received at the expense of the plaintiff? and
(iii) Was the enrichment received in circumstances of injustice?

The main difficulty lies in interpreting in (iii) above what circumstances are to be considered 'circumstances of injustice'. This is, generally, decided by the courts, and the tendency has been for the courts to only award restitution where the circumstances of the enrichment fell within one of the recognised grounds of restitution. This approach is losing favour and it has been argued recently that more judicial discretion should be used in restitution cases:

> *An action of restitution appears to me to provide an almost classic case in which the jurisdiction should be available to enable the courts to do full justice ...The seed is there, but the growth has hitherto been confined within a small area. That growth should now be permitted to spread naturally elsewhere within this newly recognised branch of law. No genetic engineering is required, only that the warm sun of judicial creativity should exercise its benign influence rather than remain hidden behind the dark clouds of legal history.*[3]

It has been ruled that where a sub-contractor had been overpaid and had then failed to complete the full extent of its works, the contractor could recover the overpayment despite having the work completed without in fact incurring any loss.[4] If the court had ruled otherwise it would in effect have supported unjust enrichment.

Another construction-related example of unjust enrichment which we are able to relate to, although different from the everyday life example given above, is where a mechanical sub-contractor, producing drawings for mechanical layouts, is pressurised by the contractor into checking its drawings against those of other sub-contractors for instances of clashes either with steel members, electrical conduits, ductwork or whatever. This is often a service which the sub-contractor starts providing either as a gesture of goodwill to help out and gain favour, to make its own job easier or through ignorance of the true extent of its responsibilities under the contract, and which progresses until it realises that it has taken on a role of far greater significance and responsibility than it had intended.

When it asks the contractor for payment it is told that it is doing no more than the contractor expected it would under the terms of the sub-contract. The contractor in this example has benefited substantially from the expertise of the sub-contractor. There is no written agreement between them stating how the additional work is to be valued, and the contractor in any case argues that the work is not additional. The sub-contractor has no recourse under the sub-contract between himself and the main contractor other than to invoke the dispute resolution procedures prescribed in the sub-contract.

If the sub-contractor could show from the practices usual in the industry or from correspondence between the parties that the clash checks were not its responsibility, it may be able to claim restitution from the contractor on a fair value basis, using the argument that the contractor has gained a benefit unjustly. This would need justification to the extent that there was an implied intention to pay and/or that the sub-contractor did not intend this work to be understood as part of its scope of work, possibly in the latter case since normal industry practice showed this to be the generally accepted rule.

I witnessed one situation similar to the above when our joint venture partner, responsible for design and construction management, sub-contracted a major part of the mechanical work, including its design, to a specialist contractor and assumed that the specialist would carry out clash checks to ensure there were no clashes between its piping and equipment and the structural steelwork, slabs, walls and so on. Unfortunately for our partner, the specialist was worldly-wise and knew that, according to accepted industry practices, this was not part of its scope of work. However the matter was not raised by either of them.

At the construction stage our general contractor, sub-contracted to my part of the joint venture (it was a typically complicated Japanese/US joint venture agreement), started submitting daywork records for additional work overcoming clashes between the mechanical and structural works. These amounted to several million US dollars. Our joint venture partner initially held that these additional costs were our responsibility since the general contractor was under our control, then when that failed it blamed the mechanical sub-contractor for not carrying out adequate checks at the design stage. When that sub-contract was analysed and found to contain no such responsibility our joint venture partner had no alternative other than to accept that it was at fault.

Their unjust enrichment would have been the saving of fees that they had anticipated through unjustly passing their responsibility on to their sub-contractor. When they tried to pass the costs on to my part of the consortium they were trying to benefit from their wrongdoing (not carrying out the clash checks themselves) by not having to bear these additional costs, or even any part of them, themselves.

The best approach to prevent these types of situations occurring is, as always, to ensure that no work is carried out without a written instruction and without the divisions of responsibility and the terms of payment being clearly agreed between the parties.

1. The legal theory behind unjust enrichment is the constructive trust, which the court imposes upon the circumstances to hold the person unjustly enriched as the trustee for the person who should properly get the property back, held to be the beneficiary of the constructive trust.
2. *Fibrosa Spolka Akcjna* v *Fairbairn Lawson Combe Barbour Ltd* [1943] AC 32; Lord Wright.
3. *Westdeutsche Landesbank Girozentrale* v *Islington LBC* (1996) AC 669.
4. *D. O. Ferguson Associates* v *Sohl* (1992) 62 BLR 95.

Restriction on Trade

Occasionally we come across situations where there is a problem providing all or part of the goods contracted for because of some government or international body having issued some form of restriction on trade between the country of origin of the goods and the country of destination. Governments may impose restrictions on trade by any of the following methods:

- Tariffs – these are import duties or taxes that people must pay when they export goods to the country imposing the tariffs. Tariffs can make imported goods more expensive, thereby encouraging people to buy goods produced at home rather than imported goods. The importing country's revenues can be increased either through generating an increase in sales of home-produced products or from the additional tariffs levied.
- Quotas – these are physical restrictions on the amount or quantity of goods that can be imported into a country over a period of time. An example occurred in the UK where the government applied a quota system to restrict the number of cars that could be imported from Japan in an attempt to protect domestic businesses from overseas imports and in so doing protect the jobs of domestic workers.
- Subsidies – by providing money for home producers the businesses in the home country can compete on price against imports from countries with lower production costs. This also helps protect against 'dumping' of foreign goods (goods which are exported at a very low price by countries wishing to establish a market in another country).
- Embargoes – this is when bans are placed on the importation of certain items from certain overseas countries. Embargoes are becoming a particular problem as governments impose them as part of their international fight against terrorism and against the proliferation of nuclear and other warfare capabilities of developing countries. There are also embargoes based upon differences in religious beliefs, in particular embargoes imposed by some Arab countries on goods of Israeli origin or provided by companies with links to Israel.

These restrictions should be addressed when preparing the original bid but nevertheless it is well worth addressing them again at the initial stages of the contract to ensure that no new restrictions have been imposed between the time of the bid and time of signing the contract which could have a significant effect on the execution of the works. If changes have been made it might be possible to claim recompense for any additional costs to be incurred complying with the new regulations but this would have to depend upon the precise wording of the relevant clauses in the contract and the precise timing of their imposition.

In the United States, President George W. Bush took the decision to impose a tariff on steel imports into the US in 2003 amounting to 30 per cent. US steel producers were initially pleased by his actions but steel users in the US were not, and were faced with a choice – either continue buying from foreign suppliers and pay a higher price or buy from the US producers. The second option, which was the option Bush was aiming at in order to protect US jobs, seemed at first glance to be the best for US steel users but this option was clouded by the major US steel producers taking advantage of the protection they had been afforded by Bush's decision – they raised their prices.

In order to find a way round the problem, some steel users rearranged their supply chain and rather than manufacturing component parts in the US from steel purchased in the US they transferred their manufacturing base to Canada. The US-imposed tariff did not affect steel imports landing in Canada and so getting a Canadian company to import steel from overseas, manufacture the component and then ship the completed component parts to the US helped to keep costs low and avoided having to pay the tariff. It did nothing, however, to protect US jobs and so had a negative effect overall on the US economy.

When faced with such a situation on a construction contract, either directly or indirectly, the question to be asked is 'what is the balance of the costs and benefits' for all the interested parties of the different options available? There is not always a simple solution and it might be that finding a source of supply from a different country is the only option available, with the result that time and cost impacts are felt by the contractor. The contractor will want to claim recompense in respect of both the time and cost from the employer but this is not always possible, depending upon the nature of the restriction, the timing of its imposition and the exact wording of the contract.

On a contract in one particularly strict Arab country, the company I was working with was involved in the fitting out of a hotel and conference centre that was being constructed for the government. The furniture we had to import included baby's high chairs for the restaurant. Our source of supply of the high chairs was proposed to our client and duly approved. The high chairs were ordered from the approved suppliers in Europe and duly arrived at the port.

The next we knew all the high chairs were going up in flames. Our supplier had re-routed the order to a company in the UK that it frequently purchased such items from. What our supplier was not aware of, however, was that the company it was using had recently established a factory in Israel and had accordingly been placed on the Arab government's blacklist of suppliers, hence all imports from the company, irrespective of where they were actually manufactured, were embargoed.

We had no time to re-order and had to find a local company that could produce the nearest equivalent to the high chairs we had originally ordered. The additional time and cost impacts were significant but unfortunately we had no clause in our contract similar to that suggested to protect us. This indicates the need not only to check restrictions at the tender stage but also at each stage of the contract, particularly when procuring items from outside the country where the work is being carried out.

Some Arab countries, including Saudi Arabia,[1] have restrictions on the percentage of the contract works that can be purchased from overseas and have also been known to insist in the contract on specific material purchase orders or sub-contract orders being placed with local manufacturers or sub-contractors.[2] Tender reviews boards take account of the percentage of local participation anticipated when deciding on the tender award.

The National Industrial Participation Programme (NIPP) of the Republic of South Africa requires that the Industrial Participation Secretariat of the Department of Trade and Industry (DTI) be notified of:

- any single contract equal to or greater than US$10m;
- multiple contracts for the same product or services, each exceeding US$3m awarded to one seller over a two-year period, the total equal to or greater than US$10m; and
- a contract with a renewable option clause where, should the option be executed, the US$10m limit would be equalled or exceeded.

This became obligatory on 1 September 1996 (fully endorsed by Cabinet approval on 30 April 1997) and in effect means that all government and parastatal[3] purchases or lease contracts (goods, equipment or services) with an imported content equal to or exceeding USD 10 million (or the equivalent thereof) are subject to an Industrial Participation Obligation. The obligation is upon the supplier, and the supplier enters into an agreement with the government, not with the purchaser. The objectives of the NIPP are stated to be:

- enhancement of sustainable economic growth;
- establishment of new trading partners;
- encouragement of foreign investment in South Africa;
- an increase in exports of South African value added goods and services;
- research and development collaboration in South Africa;
- job creation;
- human resource development;
- technology transfer;
- economic advantages for disadvantaged communities.

Principles included in the NIPP procedures are:

- the Industrial Participation obligation is 30 per cent of the imported value of the purchase contract;
- the Industrial Participation obligation must not result in an increase in the price of the purchase;
- the Industrial Participation must be profitable to the seller and beneficial for the South African economy;
- all Industrial Participation proposals must reflect incremental or new business

to be considered for Industrial Participation credits;
- all Industrial Participation projects must be economically and operationally sustainable, even after the discharge period.

Each potential supplier has to submit a business concept to the IP Secretariat followed, upon approval of the concept, by a business plan. If this is approved the IP Secretariat notifies the Industrial Participation Control Committee and if they agree the supplier then enters into an Industrial Participation Agreement with the DTI.

It is suggested that a clause should be inserted in all contracts to protect the contractor from risk in case a restriction not allowed for at the tender stage should be imposed during the course of the contract. If no such clause is included then the contractor will possibly be able to rely upon the force majeure clause should such a situation arise and would be wise to ensure at the pre-contract stage that the wording of the force majeure clause gives sufficient protection in this respect. The restriction on trade clause should:

- empower the contractor to claim reimbursement of any additional time or costs incurred as a result of the imposition of a restriction after the date the contract was signed;
- spell out the range of restrictions in order to avoid later confusion and should include:
  - international regulations and/or restrictions on trade with countries of temporary and/or final destination (such as UN resolutions);
  - laws, statutes, rules and regulations of the country of origin restricting trade with the countries of temporary and/or final destination;
  - laws, statutes, rules and regulations of the country of temporary and/or final destination restricting trade with the country of origin;
  - internal codes, rules or regulations of the contractor and/or its suppliers that similarly restrict its scope of supply to the countries of temporary and/or final destination.

1. Saudi Arabia's Council of Ministers' Resolution 124 requires that foreign contractors awarded government projects must sub-contract at least 30 per cent of the value of the contract to wholly-owned Saudi sub-contractors unless they can show that no Saudi sub-contractor can provide the required goods or services. Foreign companies with a majority Saudi shareholding are exempt from this Resolution.
2. Details of the restrictions currently in force can be obtained from the Saudi Ministry of Commerce and Industry's Directorate General of Supply.
3. A 'parastatal' is a body corporate set up by an Act of Parliament with specific objectives within the general framework of the Government's planned economic and social development programme. The main aim behind establishing a parastatal body outside the direct control of central Government machinery was originally to allow the body a higher degree of autonomy and freedom from cumbersome bureaucratic procedures to facilitate decision taking, thus ensuring timely achievement of planned objectives.

Retention

The withholding of retention by employers on contractors and by contractors on their sub-contractors and suppliers is becoming a cause for concern throughout the construction industry. There is a growing campaign amongst the industry's representative bodies to abolish this practice, but it is doubtful whether they will achieve any positive results to their campaigns, at least not in the immediate future.

It is far more likely that over the coming years industry practices will gradually change to make this less of an issue to contractors and, to a greater extent, to sub-contractors and suppliers. These latter two classes are generally the worst affected by this practice since their conditions are usually back-to-back with the main contract provisions between the contractor and the employer.

In the normal course of events employers deduct around five per cent of the total value of each payment under the contract from each interim or milestone payment, up to a limit of five per cent of the contract sum (these percentages can vary depending upon the particular terms of the contract concerned) and hold this throughout the term of the project. Again in the normal course of events this figure is reduced by 50 per cent when the project is completed (i.e. when the Certificate of Practical Completion or Taking Over Certificate is issued) and the balance of 50 per cent is released when the Final Certificate is issued (upon completion of the Defects Liability Period, which is usually 12 months after Practical Completion). Main contractors invariably pass this requirement down to their sub-contractors and suppliers.

It is possible for a contractor to take out a Retention Bond,[1] which enables the employer to make interim or milestone payments to the contractor in full, without any retention deductions, upon the receipt of a bond guaranteeing repayment of the retention monies covered by the bond upon the contractor's default in fulfilling his defects and warranty obligations under the contract. This is also open to sub-contractors but is not always a viable option for them.

Employers like the idea of holding retention as it give them some degree of comfort in case of the contractor's insolvency or its failure to complete remedial and replacement works necessary during the defects and warranty periods. It is estimated that in the UK alone at any given time there is a vast sum in excess of $4.5bn locked up in retentions in the construction industry. Much of this is held on sub-contractors and suppliers who are at a disadvantage to main contractors. The whole process does immense harm to the competitiveness and viability of the small and medium-sized companies in particular.

It is somewhat strange to note that the principal standard forms of contract make no provision for retention monies to be held in a dedicated neutral bank account or in trust, thus protected from the insolvency of the employer.[2] As it stands the employer can use the retention funds for whatsoever purpose it proposes and the contractor has no means of stopping it from doing so.

Sub-contractors have their own concerns, the main one being that the main contractor may not fulfil its defect and warranty obligations in due time, if at all. The defect and warranty items concerned may have no bearing upon the sub-contractor's work and the sub-contractor may well, in fact, have completed all of its own defect and warranty obligations in due time yet still have retention withheld from its account. The sub-contractor also faces the possibility of not only the employer but also the main contractor becoming insolvent, thus negating its chances of ever recovering its retention monies in full. The sub-contractor can be excused for regarding this as an unjust situation.

It is a well-known rule of law that a party cannot benefit from its own breach. If the contractor has not carried out its defect and warranty obligations under the main contract then it is in breach and will not receive its outstanding retention monies from the employer. By not paying the sub-contractors their retention monies it could be argued that the contractor is benefiting from its breach and that it cannot escape its liability to its sub-contractors in this way.

Until quality and efficiency are improved in the construction industry it is hard to see how the present situation can change. The feelings of distrust held by the employers (fear that the contractor will not remain solvent and that it will not complete either the works or its defects and warranty obligations) are countered by the contractor's feelings of distrust (fear that the employer will keep finding dubious faults with the contractor's discharge of its obligations as a strategic means of delaying release of retention funds, or that the employer will in fact itself sink into insolvency, taking the contractor's retention monies down with it).

Until the whole atmosphere of the industry changes there are not likely to be any changes in the withholding of retentions.

1. See 'Bonds and Guarantees' section on p.29.
2. There is no reason why a contractor, or sub-contractor, could not ask for a clause to be written into the main contract, or sub-contract as the case may be, to the effect that all retention monies will be lodged in a dedicated bank account and be subject to particular withdrawal procedures in case of the employer's insolvency.

Rights of Third Parties

England has long been one of the countries that have adopted the privity of contract[1] principle in that no party other than one of the parties to the contract can be sued or sue under it. This all changed with the introduction of the Contracts (Rights of Third Parties) Act 1999. Similar Acts have been introduced in Hong Kong, New Zealand, Singapore and other countries.

Under the provisions of these new Acts, a third party may enforce a contractual term where the contract either expressly provides for that right or where the contract confers or purports to confer such a right on it, provided that if on a proper interpretation of the contract it appears that the parties did not intend the term to be enforceable by a third party, no such right would apply. For a third party to be able to enforce the rights granted to it under the contract, that third party must be identified in it, either individually or as a member of a specific class or group or answering a particular description, although the third party does not actually have to be in existence at the time the contract was entered into. This could result in a contract conferring rights on all employees, present and future, all purchasers of products and even businesses not yet established.

The third party would have all the remedies available in a breach of contract situation that would have been available to it had it been a party to the contract, and could claim damages, obtain specific performance or injunctive relief. The third party would also have to abide by the general contract rules in respect of mitigation, remoteness of damages and so on but it could not terminate the contract for breach. The third party would be able to use any exclusion of limitation clause in the same way it would have been able to had it been a party to the contract and would, in a dispute situation, be deemed a party to the arbitration agreement contained within the contract.

Damages should compensate the third party for damages, loss or injury caused by the subject breach and should be such as would put the third party in the position it would have been in had the breach not occurred. The damages suffered by the third party must have arisen ordinarily and naturally from the breach and must have been reasonably foreseeable by the parties at the time that the contract was entered into.

The other points of interest in international contracts is that the third party rights do not apply in respect of:

- customers of an employer enforcing a contractual term in an employment contract against an employee;
- contracts on a bill of exchange, promissory note or other negotiable instrument;
- contracts for the carriage of goods by sea and appropriate international transport conventions;
- Articles of Association of a company and the rights of its members.

Acts in other Common law countries such Singapore are basically the same as the Act of England and Wales. The Singapore Act has two principal differences taken from the New Zealand Contracts (Privity) Act 1982:

- the third party's rights to equitable remedies will not be defeated merely by the maxim that 'equity does not aid a volunteer'; and
- the third party need not have knowledge of the precise terms of the contract in order to invoke the bar on the other parties' rights to vary or rescind the contract by agreement, as long as it has actually relied on the term with the parties' knowledge or in circumstances where it was reasonably foreseeable.

The position under Civil law jurisdictions is not that different from the situation under Common law jurisdictions. The French Civil Code equivalent of the English doctrine of privity of contract states:[2]

> *Agreements produce effect only between the contracting parties; they do not harm a third party, and they benefit him only in the case provided for in Article 1121.*[3]

Article 1121 noted in Article 1165 quoted above states:

> *One may likewise stipulate for the benefit of a third party, where it is the condition of a stipulation which one makes for oneself or of a gift which one makes for another. He who made that stipulation may no longer revoke it, where the third party declares that he wishes to take advantage of it.*

Both Common law and Civil Code systems are therefore driven by the same basic principles, and each have been forced by events to make certain exceptions. The principal French Civil Code exception as stated in Article 1121 of the Code, which occurs when the parties to a contract intend that rights or duties created by that contract should attach to a third party, seems to have been written as a qualification to Articles 1119 and 1120 which state:

> *Article 1119: As a rule, one may bind oneself and stipulate in his own name, only for oneself.*
>
> *Article 1120: One may, nevertheless, stand as guarantee for a third party, by promising his acting; subject to a compensation against him who stood guarantee or promised to obtain ratifications, where the third party refuses to keep the undertaking.*

There is not, however, any exception under the French Civil Code (unless we consider Agency and Assignment to be exceptions) to the principle that a contract cannot impose a burden upon a third party.

The German Civil Code states:[4]

> *An obligation with duties in accordance with [Article] 241(2) may also arise towards people who are not intended to be parties to the contract. Such an obligation arises in particular if the third party by enlisting a particularly high degree of reliance materially*

influences the contractual negotiations or the conclusion of the contract.

Article 241(2) of the German Civil Code referred to in the last quotation states:

An obligation may require each party to have regard to the other party's rights, legally protected interests and other interests.

The majority of internationally-accepted standard contracts already include exclusion clauses excluding the provisions of these relatively recent Acts from their contracts. It is advisable that a clause be inserted in every bespoke contract in the interests of risk protection.[5]

1. The doctrine of 'privity of contract' has been explained as meaning a person cannot acquire rights or be subjected to liabilities arising under a contract to which it is not a party. However, such a contract may still affect third parties in respect of the creation of proprietary or quasi-proprietary rights or by giving rise to a duty of care to a third party. A contract can also be said to impose on a third party the duty not to interfere with the contracting parties in their performance of the contract they have entered into.
2. French Civil Code, Article 1165.
3. *ibid.*, Article 1121.
4. German Civil Code, Article 311(3).
5. A typical clause would be worded: 'Nothing in this Contract confers or purports to confer any third party benefit or right to enforce any term of this contract pursuant to the Contract (Rights of Third Parties) Act 1999 except and to the extent that this Contract expressly provides for such Act to apply to any of its terms.' In countries other than England or Wales the Act specific to that country would be substituted for the one stated in this typical clause. (for example, in Singapore we would refer to the 'Contracts (Rights of Third Parties) Act 2001').

Risk Allocation, Assignment and Assumption

With the whole concept of construction project risk changing over the past decade or two as the number of unknown and unpredictable problems has increased, so the risk management terminology has grown to include such terms as 'risk allocation', 'risk assignment' and 'risk assumption'. The significant changes in risk categories and the approaches to minimising their effects has been brought about to a large extent by the different contract relationships that have developed, including partnering, consortia and joint ventures. The whole contract environment has developed of necessity to meet modern demands and to maximise the benefits of capital expended.

There are several stages to go through in the process of reviewing the potential risks to a project. These can best be defined as:

- establish the context:
  - strategic – bid stage, negotiation stage or post-contract stage;
  - organisational – client, consortium, sub-contractor/supplier;
  - risk management – technical, commercial (cost), commercial (time), political;
- decide the structure:
  - term or condition;
  - change order;
  - other;
- identify the risk:
  - what can happen?
  - how can it happen?
- analyse the risk:
  - determine likelihood;
  - determine consequences;
- evaluate the level of risk;
- evaluate the risk:
  - compare against criteria;
  - set risk priorities;
- set strategies:
  - identify strategic options;
  - evaluate strategic options;
  - prepare strategic plans;
- implement strategic plans.

The general levels of risk that are encountered in an international contract context can be categorised as follows, although it must be noted that several of these risks overlap in practice:

- physical;
- ability related;
- cost related;
- time related;
- design, engineering and construction related.

Physical risks would include events or occurrences that prevent or delay physical completion of any element of the works. Examples would be:

- force majeure events such as earthquakes, tidal waves etc.;
- site access – physical obstruction;
- impossibility of performance;
- site conditions (particularly sub-surface conditions);
- quantity increases and decreases;
- exceptionally inclement weather (possibly included in force majeure clause).

Ability related risks would be events or occurrences that prevent or delay completion of the works other than physical events and occurrences. Examples would be:

- defects in the work;
- inadequate labour;
- sub-contractors;
- contractor-supplied material/equipment shortages/delays;
- strikes or lockouts (possibly covered by force majeure clause).

Cost related risks would be those risks that could result in changes to the contractor's project costs, and hence affect his profit margin. Examples would be:

- performance and other relevant bonds;
- changes in laws and regulations;
- escalation of costs;
- exchange rate fluctuations;
- devaluation;
- fire, theft, accident etc.;
- termination;
- payment issues;
- insurance;
- funding;
- taxes;
- warranties.

Time related risks are those events that prevent the contractor completing the works within the period for completion allocated in the contract. Examples would be:

- access restrictions imposed by employer;
- employer's programme changes (particularly directed or implied acceleration);
- delays and disruptions;
- suspension;
- late approvals;
- early use of facility by employer.

Design, engineering and construction related risks are other events and occurrences that affect the contractor's performance in terms of time and/or money. Examples would be:

- changes to extent, manner of execution or cost of the works;
- continuation of work during dispute;
- coordination of sub-contractors and suppliers and other contractors on site;
- guaranteed performance values;
- errors in documents provided by employer;
- ambiguities and inconsistencies in and interpretation of contract provisions;
- intellectual property/copyright;
- method of construction;
- employer-supplied materials and equipment;
- permits and licences;
- productivity;
- responsibility for care of the works;
- site safety;
- quality control.

Table 1 should give an idea as to how these various risks may be allocated under an international construction contract. There are many different approaches as between the FIDIC Red, Yellow and Silver Books, the US EJCDC forms of contract and other specialist industry forms of contract so Table 1 must be viewed as a guide only. A similar table should be prepared for each project in order to identify the allocation of risk in a form that the whole project team can easily relate to. This would then serve as a useful tool in the risk management process.

Table 1: Allocation of risks under international construction contract

Risk Description	Employer's Risk	Contractor's Risk
Physical Risks:		
Force Majeure Events	Partial – waives liquidated damages and grants the contractor a non-compensatory time extension	Partial – obtains non-compensatory time extension, so bears additional overhead and other cost risks
Site Access for Construction	Yes	Can claim time/cost as damages
Impossibility of Performance	Yes in respect of planning, zoning and similar permissions and buildability of design (except where contractor's design)	Yes in respect of buildability of design under FIDIC Silver Book and similar Design & Build contracts, except where due to employer's interference
Site Conditions	Yes where sub-surface or physical conditions materially differ from those included in employer's documents, unless otherwise stated	Yes under FIDIC Silver Book and similar Design & Build contracts. Where employer's risk, contractor entitled to additional time and cost
Quantity Increases/ Decreases	Yes under re-measurable contract	Yes under non-re-measurable lump sum contract
Ability Related Risks:		
Defects in the Work	n/a	Yes
Inadequate Labour	n/a	Yes
Sub-Contractors' Performance	n/a	Yes
Shortages in Contractor-Supplied Materials and Equipment	n/a	Yes
Cost Related Risks:		
Performance and Other Bonds	n/a	Yes
Changes in Laws	Yes if changes occur after the base date of the Tender	n/a
Escalation of Costs	n/a	Yes unless specific escalation clause states otherwise
Exchange Rate Fluctuations	n/a	Yes unless specific clause states otherwise, or unless due to change in law (devaluation might be one such case depending on particular circumstances)

Risk Description	Employer's Risk	Contractor's Risk
Fire, Theft, Accidents	Yes where indemnity given in respect of own employees etc., passed on to insurance provider	Yes, passed on to insurance provider
Termination	Yes for convenience termination; Yes for failure to pay or for long-term stoppage	Yes for contractor's failure to perform
Payment Issues	Yes – must ensure payment claims are accurate and timely to avoid risk of overpayment	Contractor at risk from Employer's late and non-payment affecting its ability to pay its sub-contractors and suppliers and to recover its own costs
Insurance	Partial	Partial
Funding	Yes	Yes where contractor provides all or part of finance as part of his contract
Taxes	Yes to the extent that specific clause gives responsibility for tax changes and increases after contract date	Yes in respect of tax changes and increases not covered by specific clause rendering them reimbursable
Warranties	n/a	Yes, partially passed on to sub-contractors and suppliers
Time Related Risks:		
Access Restrictions Imposed by Employer	Yes	Entitled to paid time extension
Employer's Programme Changes	Yes unless programme change due to Contractor's delay	Entitled to paid time extension except where due to own delay
Delays and Disruptions	Yes if employer's fault – waive liquidated damages and grant paid time extension to contractor	Yes if contractor's fault – liable for liquidated damages and mitigation costs
Suspension	Yes	n/a
Late Approval of Timely Submissions	Yes	n/a
Early Use of Facility by Owner	Yes	n/a
Design, Engineering and Construction Related Risks:		
Changes to Extent, Manner of Execution or Costs of the Works	Yes if initiated by employer	Yes if initiated by contractor for own convenience
Continuation of Work During Disputes	n/a	Yes

Risk Description	Employer's Risk	Contractor's Risk
Coordination of Others	Yes in respect of other contractors employed by employer on same site. Often employer tries to pass this risk on to contractor	Yes in respect of own sub-contractors, suppliers
Guaranteed Performance Values	n/a	Yes
Errors in Employer's Documents	Yes	In FIDIC Silver Book this risk is passed on to contractor
Ambiguities and Inconsistencies in and Interpretation of Contract	Yes	n/a
Method of Construction	n/a	Yes
Owner-Supplied Materials and Equipment	Yes	n/a
Permits and Licences	n/a	Yes, subject to employer giving reasonable assistance
Productivity	n/a	Yes, unless interfered with by employer
Care of the Works	n/a	Yes
Site Safety	n/a	Yes
Quality Control	n/a	Yes
Setting Out	Yes as far as accuracy of items of reference provided by employer is concerned – in such case contractor is due a paid extension if delayed	FIDIC Silver Book passes risk on to contractor

Ruling Language

In the international construction field it is commonplace for a contract to be entered into that requires a contractor from one country to work for a client in another country and even to actually execute the works in a totally different country. It is easy therefore to understand that the ruling language of the contract can be an extremely important issue. The ruling language should be clearly defined in the contract documents and there should be no room for misunderstanding as to its application.[1]

The language in which the management of the contract is to take place, including day-to-day correspondence, minutes of meetings and conversations between client and contractor employees, may be different from the ruling language of the contract and if this is the intention then this too should be clearly defined within the contract documents. The contract documents should also clearly define the language in which arbitration proceedings would take place. Managers should be fully aware of the ruling language to be applied in each particular circumstance.

The contract documents themselves may have been initially drawn up in the language of the country of the employer and then translated into another language in order to assist in the management of the bidding and construction processes. It is often found in such cases that, in case of ambiguity, lack of clarity or other dispute the original documents written in the language of the country or employer will be given preference over the translation made to assist in day-to-day management. This again must be clearly defined in the contract documents.

Contractors will be aware of which languages lend themselves to translation into, for example, English and which do not. In the latter case the contractor would be wise to employ an interpreter to review the translation and to identify translated items that could be the cause of later disputes.

Some countries and employers require all correspondence to be in more than one language[2] and in such a case it would be essential to procure the services of a highly trained and reputable translator familiar with the construction industry. As well as translating the contractor's correspondence to the employer the translator would also be able to comment on any inaccuracies or ambiguities in the employer's translation from the original to the administrative language. This is especially important with regard to instructions, notices and claims where a slight ambiguity in translation could have extensive ramifications in terms of time and/or cost.

I remember reading about a contract between a UK employer and a French contractor, written in both English and French texts, which stated in the English text that the law of the contract was to be the law of the 'place of execution of the contract'. The English definition of 'execution' of a contract in a legal context is the act of 'signing, sealing and delivery'[3] of the contract, whereas the French had translated 'execution' into what would in English be the equivalent of 'performance'. The project was built in Algeria whereas the contract was signed in England, giving rise to great confusion as to what was actually the law of the contract – either the law of England where the act of signing, sealing and delivery of the contract took place or Algeria where the subject matter of the contract was being performed. Fortunately the contract documents contained a clause that stated the language of the contract was English and, in cases of conflicts arising from translations, the English version would take precedence.

1. It would be sufficient in most circumstances for such a clause to say: 'The ruling language to apply to the Contract shall be English and the Contract shall be construed, interpreted and administered, and all correspondence between the Parties shall be in, that language.'
2. In Saudi Arabia, for instance, contracts are frequently written in English and Arabic but in all cases of dispute the Arabic text will prevail. All business with government bodies must be in Arabic. Also in Saudi Arabia all Joint Venture Agreements or Agency Agreements must be in the Arabic language to be legally binding.
3. See *Osborne's Concise Law Dictionary* – 'Execution' paragraph (2).

Set-Off

The construction industry throughout the world attaches great significance to the right of set-off in the administration of construction contracts. It is commonplace for interim payments from contractors to sub-contractors to be reduced from that otherwise due by an amount which represents contra-charges (the financial means of applying a set-off) for services provided by the main contractor (such as electricity, water, security services and so on) and for agreed damages that the main contractor has incurred as a result of some breach, action or lack of action by the sub-contractor.

The basic principle of the right of set-off enables a debtor to reduce or extinguish a creditor's claim by the amount of any cross-claim which the debtor might have against the creditor. This procedure is, unfortunately, often abused, both by the employer and the contractor where the parties each take advantage of the contra-charge system to reduce payments (employer to contractor and contractor to sub-contractor) in a cavalier fashion, sometimes without real justification.

There are four basic types of set-off:

- legal set-off;
- equitable set-off;
- banker's set-off;
- insolvency set-off.

Legal set-off can only be resorted to as a defence to a court action and the cross-claims concerned must be in respect of liquidated debts or money demands which can be proved with certainty. The debts must be:

- both due and payable at the time of the pleading;
- between the same parties; and
- held in the same capacity, right or interest (satisfying the test of mutuality, which would not be satisfied, for instance, if the set-off debt was owed by a party in its individual capacity whereas the claimant's debt was a corporate debt).

Legal set-off debts do not necessarily have to result from the same or even closely related transactions. Cross claims which are only estimates at the time that the set-off is proposed and can only be ascertained with finality by litigation or arbitration would not be accepted as valid claims for set-off.[1]

Equitable set-off differs from legal set-off in that:

- it does not need litigation to make it an available debtor option; and
- the cross claims must arise from the same or closely related transactions.

A debtor can simply deduct the amount of its alleged cross claim from debts otherwise due to the claimant and merely pay the balance, if any. Where the transactions were related to different contracts but were part of a single trading agreement this would not necessarily be regarded as a sufficiently close relationship to justify equitable set-off.[2] Unlike legal set-off, claims may be for un-liquidated sums, such as damages, and mutuality is not a pre-requisite, although it would normally be considered inequitable or unjust for the right of set-off to be applied between parties where mutuality was not present.

The sums concerned must again be both due and payable at the time of the set-off. If the set-off is in respect of un-liquidated damages it must be based upon a reasonable assessment of the loss suffered and must be made in good faith.[3] It is not possible to claim a right of set-off merely because cross claims exist; on the contrary, it is necessary to show that the debtor's cross claim is so closely related to the claimant's demand that it would be unjust to enforce payment of the claimant's claim without taking due account of such cross claim.[4]

Banker's set-off is a banker's remedy where a bank's customer has more than one current or loan account with the same bank and one of them falls behind in a debt repayment. The bank may in such a case transfer monies from the account in credit and use them to pay off the debt in the other. This is a similar remedy to equitable set-off in that no litigation is necessary in order to make its application possible.

Insolvency set-off is covered by mandatory rules of insolvency contained within the Insolvency Act 1986 applicable in England and Wales and in similar statutes in other countries. Insolvency set-off cannot be varied by terms written into the contract. There are only a few narrow circumstances in which set-off is allowed during a bankruptcy since in principle it goes against the general rule that all non-priority unsecured creditors are to be treated equally when settling their claims. In the US a distinction is made in insolvency actions between set-off and recoupment[5] in that the right of set-off may be applied where the cross claim arises out of different transactions whereas in recoupment the debts must arise out of the same transaction. In other set-off situations there is virtually no distinction between set off and recoupment.

In general, and with the exception of insolvency set-off, it is acceptable to include a clause within a contract which restricts[6] or excludes the right of set-off between the parties. Such clauses are enforceable, if properly drafted, and are not considered to be contrary to public policy, but the wording must be clear and unambiguous. Under English Common law such a clause is treated as a limitation of liability clause and is thereby subject to scrutiny in relationship to the reasonableness test included in the Unfair Contract Terms Act 1977 and also in relationship to the Unfair Terms in Consumer Contracts Regulations 1999.

Under the German Civil Code:[7]

> *the fact that a claim is time barred does not preclude set off and the assertion of a right of retention if the claim was not time barred at the moment when set off could first have been made or performance refused.*

Also under the German Civil Code[8] even where derogation from the statutory provisions is permissible, several clauses are noted as being invalid if used in standard business terms including:

> *3. (Prohibition of set off)*
>
> *A provision by which the contractual partner of the user is deprived of the right to set off a claim which is undisputed or has been declared final and absolute.*

Under the French Civil Code:[9]

> *Set-off is brought about as of right by the sole operation of the law, even without the knowledge of the debtors; the two debts are reciprocally extinguished, from the moment when they happen to exist at the same time, to the extent of their respective amounts.*

In addition under the French Civil Code:[10]

> *Set-off takes place only between two debts which have likewise as their object a sum of money or a certain quantity of fungibles of the same kind and which are likewise liquid and due.*

1. *B. Hargreaves Ltd* v *Action 2000 Ltd* (1992), CA.
2. *Bim Kemi* v *Blackburn Chemicals Ltd* (3 April 2001), CA.
3. *The Nanfri* (1978) 1 QB 927.
4. *A. B. Contractors Ltd* v *Flaherty Brothers Ltd* (1978) CA.
5. Keeping or holding something back that would otherwise be due because there is a fair and just reason for doing so.
6. A typical clause in an international construction contract would empower, but not obligate, either party to apply set-off provided that such set-off amounts had been agreed to by the other party. Unlimited powers to apply set-off where no agreement had been reached on the amounts concerned would be considered unfair.
7. German Civil Code Section 215 of the Act to Modernise the Law of Obligations effective 1. January 2002.
8. *ibid.*, Section 309.
9. French Civil Code, Article 1290.
10. *ibid.*, Article 1291.

Subject to Contract

The words 'subject to contract' are used in negotiation situations so as to make sure that a binding contract does not come into existence by accident during the negotiations but only after all the words have been agreed and all the parties have signed the written agreement at the conclusion of the negotiations. This is the situation in most negotiations identified as 'subject to contract'; however, depending on what actions you take during the negotiations, you may still find that a contract can come into existence other than on everyone signing the full written terms of the agreement reached.

It is important to register at the outset whether negotiations between two parties are subject to contract in order to prevent disputes should work commence without formal contracts having been executed, or should one of the parties pull out of the negotiations.

In *DMA Financial Solutions Ltd* v *BaaN UK Limited*,[1] DMA entered into negotiations with the employer, BaaN, to provide support and training services, yet at no time during the negotiations did either party register the fact that the negotiations were proceeding 'subject to contract'. In the event BaaN broke off negotiations with the specialist provider before a formal written contract had been entered into. The intention of the parties had been clear throughout the negotiations that such a formal contract would be entered into, but although all commercial terms had been agreed there were still some matters as then unresolved.

The court looked at three possible scenarios:

(i) negotiations had taken place on the basis that the parties were not legally bound until a formal written contract had been executed by both parties, even if negotiations were concluded;
(ii) negotiations were such that parties might become bound contractually even before negotiations were concluded should negotiations have reached a sufficiently advanced stage;
(iii) negotiations were as (ii) above but in the event did not reach a stage of sufficient agreement for the parties to become contractually bound.

The court held in favour of option (ii), i.e. that the negotiations had reached a stage where a binding contract had been created with all the commercial terms having been agreed. It was held that the fact that DMA and BaaN had not entered into a formal written contract did not preclude the existence of a contract at the conclusion of negotiations even though there were some matters that they had still to agree upon.

Had there been a 'subject to contract' proviso in the correspondence or negotiation records then the court would not have held so.

In a more recent case involving Confetti Records and Warner Music,[2] Confetti were approached by Warner who asked if they could include a track, the rights to which were owned by Confetti, on a remix compilation album. Confetti agreed and Warner sent them a 'deal memo' setting out the basic terms of the agreement for Confetti's signature. The deal memo was marked 'Subject to Contract'.

Confetti signed the deal memo and returned it to Warner. A management company associated with Confetti then sent Warner an invoice and a recording of the track. Time passed and the invoice remained unpaid despite Confetti writing to Warner giving a deadline for payment after which the licence to use the track would be revoked. Nothing happened and so Confetti again wrote to Warner stating that the track should be excluded from the album unless a new agreement was reached.

Warner replied that it had not received Confetti's earlier letters threatening to revoke the licence if payment was not received by the stated date and confirmed that the track had already been included on the album and 30,000 copies manufactured. The court had to decide:

- whether a binding contract was created by signature of the deal memo;
- if not, whether the sending of the copy of the track together with the invoice amounted to an offer which Warner Music was able to turn into a binding contract by making use of the track;
- irrespective of whether or not there was a binding contract, whether it was fair to allow Confetti to go back on its offer to grant a licence on the terms that had been agreed.

The judge found in favour of Warner, stating that:

> *The term 'subject to contract' had a special meaning in law. It was used to stop contracts coming into existence until a formal written contract was signed. The judge was satisfied that the music industry was no different from any other in its use of the term and as a result the signed 'deal memo' did not create a binding contract.*

However, by invoicing for the £1,500 and by sending the copy of the track for Warner Music to use, Confetti was essentially making Warner Music an offer of terms on which it would be allowed to use the track. That offer remained open until it was accepted (thereby creating a contract) or until it lapsed or was revoked.

From the evidence, it appeared that Warner Music had not received the notice from Confetti threatening revocation of the licence. By the time Confetti contacted Warner Music subsequently to revoke the licence, Warner Music had already gone to the expense of including the track on the album and of pressing 30,000 copies of it.

As a result, a contract had come into existence (having been accepted by the conduct of Warner Music in including the track and printing copies). Also, because Warner Music had relied upon Confetti's conduct (i.e. not just by Confetti signing the 'deal memo' but also Confetti sending the recording and invoice) Confetti were prevented from denying the existence of the licence.

It is easy to draw analogies in a construction context, emphasising the need to adhere to the principles of 'subject to contract' negotiations once these have been entered into and to avoid doing anything that might prejudice the protection that this endows on the negotiations.

1. *DMA Financial Solutions Ltd* v *Baan UK Limited and Others* (28 March 2000), Chancery Division.
2. *Confetti Records and Others* v *Warner Music UK Ltd* (2003) EWHC 1274.

Subrogation

Subrogation is the act of assuming the legal rights of a person for whom expenses or a debt has been paid. Subrogation is based on the principles of equity and fairness, and its goal is to place the burden of a debt upon the party who should bear it, i.e. the responsible party.

In construction contracts subrogation occurs, for instance, when an insurance company pays its client for damages, injuries and losses suffered and then sues the party that its client, the injured party, contends caused the damages. As an example, a contractor has a fire on his construction project site caused by faulty workmanship. This gets out of control and spreads to an adjacent building. The adjacent owner's insurance company which has insured the property damaged by fire pays the adjacent owner the estimated cost of reconstruction of the property, and then sues the contractor and its design team for that amount.

The insurance company will take the place of (be subrogated to) the adjacent owner in claiming against the contractor, and in turn the contractor will have all the 'defences' to the insurance company's suit that it would have had against the adjacent owner had the insurance company not been involved, including the contention that the cost of repairing the property was less than the adjacent owner was paid by the insurance company, or that the adjacent owner somehow negligently interfered with the fire fighting efforts, thus increasing the damage and the resulting cost unreasonably. The insurance company in this case would 'stand in the shoes' of its client.

Insurance companies reduce their net exposure by adopting the principle of subrogation. The insured parties also benefit by commensurate reductions in the claimed amounts which are attached to their claim record held by their insurance company resulting in stabilisation of premiums and, in some cases, by having their deductible and any other related costs incurred included as part of the insurance company's subrogated claim to the third party, to be returned to them upon success of the insurance company's action. In that way subrogation benefits both the insured and the insurer.

It is common in construction contracts to see the term 'Waiver of Subrogation'. Waivers of subrogation are clauses in the insurance requirements of business contracts that prevent one party's insurer from subrogating against (standing in that party's shoes and suing) the other party to recover a loss it has paid out. Such clauses can greatly reduce risks by identifying which party is supposed to insure the risk and prohibiting that party's insurer from trying to circumvent the intent by suing the other party for causing or contributing to the property loss. Waivers of subrogation have long been used in real estate property leases to allocate responsibility for insuring the leased property to either the tenant or landlord.

In construction contracts the waiver of subrogation should specify which party is responsible for insuring the project and waive subrogation against all other parties involved with constructing the project. Subrogation waivers may be used in all areas of insurance – general liability, contractual liability, all risks, worker's compensation and so on. – to transfer risk from one party to another. The ability of parties to contractually transfer risk is profoundly affected by legal doctrines and principles.

Whenever we are considering a contract that requires one party to purchase some form of insurance, we should consider whether it is advisable for the insurer's right to subrogate against the other party to be waived. The contract and waiver should be thoroughly and professionally evaluated to ensure that the language does not controvert the parties' intent.

In some instances, contracts may contain boilerplate[1] language that may or may not accomplish the transfer of risk desired by the contracting parties. It is important to take advice when introducing such a clause into a contract to ascertain whether the waiver of subrogation language accomplishes the desired intent, and whether it protects from unintentional assumption of risk exposure.

An insurer cannot seek subrogation against its own insured and it is wise for architects and engineers to require that the project's insurance policies name them as additional insureds, at least on the general contractor's general liability policy, all risks policy and other relevant policies. Requiring that these policies name the architect and engineer as additional insureds provides protection against subrogation claims because they then become insured parties under the policy.

Waivers of subrogation are yet another reason to use standard form agreements, at least as a starting point for our agreements, since the standard forms contain provisions that have been legally tested. In any event, waivers of subrogation are important tools for risk allocation and should be included with agreements and general conditions whenever possible.

An insurance carrier may reserve the 'right of subrogation' in the event of a loss. This means that the company may choose to take action to recover the amount of a claim paid to a covered insured if the loss was caused by a third party. After expenses, the amount recovered must be divided proportionately with the insured to cover any deductible for which the insured was responsible.

1. 'Boilerplate' language is standardised, recurring language rather than language specially tailored to suit a particular situation.

Sub-Soil Investigations Included with Tender Enquiry

Tender enquiries frequently include sub-soil reports that the employer has commissioned either directly or through its architect or engineer in order for it to progress its basic feasibility studies and initial outline design procedures. It is common for tender enquiry documentation to include exclusion clauses stating that the employer accepts no responsibility for the accuracy of such reports and advising tenderers to carry out their own inspection of the proposed site.

Where the employer is responsible for design, the courts and arbitrators tend to consider this type of exclusion clause to be inequitable and look for legal reasons to exempt the contractor from such risk. Where the contractor is responsible for design, however, he would be wise to take heed of such a clause.

FIDIC makes provision in the 1999 edition of the Red Book[1] under clause 4.12 whereby the contractor would be entitled to an extension of time 'if completion is or will be delayed', and to recover any additional costs it incurred, as a result of encountering physical conditions which were considered to have been unforeseeable at the time of tender. The ECC,[2] previously known as the NEC,[3] has a similar provision to that contained within the FIDIC Red Book 1999 edition similarly reducing the contractor's risk exposure.

The FIDIC Silver Book[4] on the other hand requires in clause 4.10 that the contractor be responsible for verifying and interpreting all sub-surface and hydrological data provided by the employer and confirms that the employer shall have no responsibility for its accuracy, sufficiency or completeness. The contractor is to accept total responsibility for having foreseen all difficulties and costs of successfully completing the works. The reason for this clause is probably due to the fact that generally in the type of turnkey project for which the Silver Book has been designed[5] the lenders rely on project costs and the project schedule not varying to any significant extent and as a result tend to insist on the contract limiting the number of possibilities for the contractor to claim extra costs or extra time to perform the works.

From an employer's standpoint it is good practice to insist that the contractor carries out its own survey. The survey costs will naturally be included in the contractor's bid costs but even if these are added to its bid price they will inevitably be less than the contingency sum which the contractor would otherwise add to its bid price to cover unexpected sub-soil conditions being discovered during the execution of the works.

Many ad hoc contracts that are not based on standard forms such as FIDIC, NEC and so on do not have any specific provision, leaving the contractor at risk. If at all possible the contractor should try to ensure that a clause is inserted into such ad hoc conditions of contract in order to minimise its risk in such a situation. Such a clause should:

- define 'unforeseeable' as anything that could not reasonably have been foreseen at the tender stage by an experienced contractor on the basis of a reasonable examination of the data relating to the works provided by the employer and of information that it could have obtained by visual inspection of the site or other data readily available at the time;

- require the contractor to notify the employer in writing of the physical conditions or artificial obstructions on the site that could not reasonably have been foreseen, the additional work required to overcome such conditions and the effects this would have on both the project schedule and cost;
- set a time limit for such notification following the discovery of the physical condition or artificial obstruction;
- empower the employer to grant an extension of time and to add the additional costs thus incurred onto the contract price.

1. Conditions of Contract for Construction for Building and Engineering Work designed by the Employer.
2. 'Engineering Construction Contract' issued by the Institution of Civil Engineers in the UK.
3. 'New Engineering Contract' issued by the Institution of Civil Engineers in the UK.
4. Conditions of Contract for EPC/Turnkey Projects.
5. The Introductory Note to the Silver Book also says clearly that the Book is not suitable 'if there is insufficient time or information for tenderers to scrutinise and check the Employer's Requirements or for them to carry out their designs, risk assessment studies and estimating (taking particular account of Sub-Clauses 4.12 and 5.1)' and further explains that 'FIDIC recommends that the Conditions of Contract for Plant and Design-Build should be used in the above circumstances for Works designed by (or on behalf of) the Contractor.'.

Tender Rate Errors

There has long been discussion on whether a tender rate that contained an error could properly be used to evaluate additional work, change orders, variations and so on.

A decision in the case of *Henry Boot* v. *Alstom*[1] changed the long-standing feeling amongst construction professionals that the rate in error is to be used up to the total related quantity in the contract bill of quantities with additional quantities being priced at a 'reasonable' rate.

In the case in question, Henry Boot were required by their employer to carry out additional works and priced them using a rate which had been included in their tender, but which was obviously in error. The relevant clause in the Henry Boot/Alstom contract provided for the use of contract rates 'as far as may be reasonable'. Alstom put forward a case that it was unreasonable to use a rate that was obviously in error.

The Court of Appeal agreed with the judge and held that the reference in the contract clause to the use of the rate being reasonable in fact meant not that the rate itself had to be reasonable but that the process of applying the rate had to be reasonable. The parties had agreed to the rate and were stuck with it. The mistake in that rate was irrelevant.

In an earlier case[2] an architect had adopted the previously held principle that the tender rate should apply to contract quantities and that a reasonable rate should be applied to quantities in excess of this. This resulted in the bills of quantities quantity of excavation in rock being priced at the low rate provided within the contract and the considerable excess quantity of rock actually excavated being priced at a rate ten times higher, this higher rate being deemed a reasonable rate by the architect. The employer had queried this and the arbitrator had found in the employer's favour. The Court of Appeal held the opinion that the financial outcome should not affect one's view on the construction of the words, the judge stating:

> *In my view the actual financial result should not affect one's view of the construction of the words. Naturally one sympathises with the contractor in the circumstances, but one must assume that he chose to take the risk of greatly under pricing an item which might not arise, whereby he lowered the tender by £1,425. He may well have thought it worthwhile to take that risk in order to increase his chances of securing the contract.*

The contractor had taken the risk of inserting a low price against an item which it thought might not be necessary but which, in the event, was found to be necessary in a big way. The contractor may well have increased its chances of winning the contract at the tender stage by inserting such an artificially low rate against the rock excavation quantity; that was a gamble that in this instance did not pay off. The Court of Appeal could see no reason for changing the arbitrator's ruling, meaning that the contract rate had to be applied to all the quantities, not just those in the contract bills of quantities.

These two cases viewed together are strong reasons for determining that the rates inserted in the contract documents are those rates which will be used in determining the value of adjustments to the works irrespective of whether those rates are in fact reasonable. This would apply whether they were unreasonably high or unreasonably low and would only apply to works of a similar nature carried out under similar circumstances, although this latter point may well be argued depending upon the exact wording of the various applicable clauses in the contract.

For instance, the low rate for rock might have been inserted against items for excavation in open cut whereas the actual rock excavated might have been in trench excavation. In such a case it would be argued that the actual works were not of a similar nature to the bills of quantities item in the contract and were not carried out under similar circumstances, since different methods of excavation would have been required, hence a new rate should be agreed upon. The counter argument to this would be to establish a 'pro rata' rate based upon the contract rate, adjusting the various elements used in building up the rate to account for the changed circumstances. The outcome of these negotiations would depend upon the actual wording of the various contract clauses, the amount by which the contract rate was adrift from that considered to be reasonable and the strength of the negotiating team.

To the many tenderers who try to second-guess what is likely to happen during the contract and who insert high rates or prices for those items of work which they expect to increase, or be the subject matter of change orders, the Henry Boot/Alstom and Dudley/Morrin cases confirmed that there is nothing reprehensible in that. Both parties enter into the contract, in theory at least, with their eyes open and must be prepared to take the rough with the smooth. As far as employers are concerned they need to be equally aware of which rates are obviously in error and to safeguard their positions accordingly.

I experienced an interesting variation of this theme when building a hotel in the Middle East. The contractor for whom I was working at the time tendered for the fitting out of the hotel and I had to prepare our bid typically within an extremely short timescale. I co-opted the assistance of my wife to type out our priced bill of quantities, which subsequently formed part of the contract. When it came to checking the bills of quantities against the actual deliveries we found that we had 120 two and a half quart casseroles whereas the contract bills of quantities listed 120 twelve and a half quart casseroles. My wife had made a simple typing error, inserting a one before the two making it twelve, and my 'thorough' checks had let this simple typing error slip by.

Anyone reading who has worked in the Middle East will understand the weeks, even months, of negotiation with the Arab inspection company working for the employer that went into resolving what was a simple mistake. It was only resolved after proving that the type of casserole required by the contract was simply not made in such a large size. Had the rate been reduced pro-rata the bills of quantities rate as the Arab consultant suggested, the final rate would have been ridiculously low – luckily common sense, and diplomacy, prevailed.

1. *Henry Boot Construction Ltd* v *Alstom Combined Cycles Ltd* (2000) BLR 247, CA.
2. *Dudley Corporation* v *Parsons and Morrin Ltd* (1959).

Time at Large

Time is said to be 'at large'where:

- there is no specific time for completion; or
- a previously fixed time for completion no longer applies.

The first case could arise where an agreement was entered into without a completion date or period being stated. This is seen in letters of intent simply instructing that work commence without stating a completion date. Where subsequent agreements do not refer to a time for completion, time will be said to be at large.

The second case could arise where:

- an act of prevention on the part of the employer creates a delay which in turn is not covered by an extension of time provision; or
- the provisions for extension of time have not been properly administered or applied; or
- there has been undue interference by the employer in the certifying process.

There is an established fundamental principle of commercial contracts which states:

> *There is a term implied into every contract by each party that he will not do anything to prevent the other party from performing a contract or to delay him in performing it. ... generally such a term is by law imported into every contract.*[1]

Most standard forms of construction contracts have built-in mechanisms, known usually as extension of time clauses, through which the employer's representative is able to extend the period for completion when delays occur through the fault of the employer or its agents. This allows a definite time for completion to be maintained and preserves the employer's right to deduct liquidated damages should completion not occur until after this extended date. The extent of the delay caused is not a relevant factor to be considered in assessing whether or not time has been set at large and even the shortest of delays could invoke such a ruling.[2]

So-called 'catch all' phrases in extension of time clauses such as 'all other causes beyond the contractor's control' do not cover interference by the employer or its representatives and in such cases an extension of time cannot be granted and time is at large.[3] In such cases express provisions are necessary to enable an extension of time to be granted, examples of such express provisions (one of which could be attached to the extension of time clause) being:

- 'Any act of prevention or breach of contract by the Employer, or any other matter in respect of which under the terms of this Contract the Employer gives the Contractor an indemnity.'
- 'Any act of prevention or breach of contract by the Employer not mentioned in this Clause.'
- '...delay caused by any act or omission of the Employer...'

In the absence of an express extension of time clause, or its application, a default of the employer that affects the contractor's ability to complete the works on time could put time at large. In *Amalgamated Building* v *Waltham Holy Cross*,[4] Denning L. J. stated:

> *... where the cause of delay is due to some act or default of the building owner, such as not giving possession of the site in due time, or ordering extras, or something of the kind ... the contract time may well cease to bind the contractors, because the building owner cannot insist on a condition if it is his own fault that the condition has not been fulfilled. That was decided in* Roberts *v* Bury Improvement Comrs *and many other cases.*

Where time has become at large the contractor's duty is to complete the works within a reasonable time. This has been defined as being a 'question of fact',[5] although this is not a particularly helpful concept and in practice each case has to be looked at on its own merits. One school of thought assesses what constitutes a reasonable time for completion after time at large by adding a reasonable time allowance to the original time plus any previously granted extensions of time for delays occurring prior to time becoming at large.

A second school of thought starts on the premise that the original time was not in any case reasonable since it was presumably agreed under a certain amount of stress in competition with other contractors. In this case the whole circumstances surrounding the project are examined and a reasonable completion date determined. Unfortunately there is little if any case law which helps promote either concept.

It should be emphasised that when time is at large the employer loses its right to deduct liquidated damages and, as noted above, the contractor's obligation is only to complete the works within a reasonable time.

It should also be noted that clauses which purport to make notice within a certain period of time a prerequisite for any claim of time at large are worthless as once time is at large the notice provisions no longer have any contractual relevance and are in effect defunct.

Where a notice is required to be issued within a set time frame under the contract and such notice is a condition precedent to an extension of time being granted, if the notice is not issued within such time frame in respect of a delay (the delay itself being 'an act of prevention' for which the employer is responsible and which would, had such notice been issued, have resulted in an extension of time being granted), applying the principles of prevention the clause in the contract under which delays by the owner would give rise to an extension of time are rendered inoperative; thus time is at large. This being the case, the owner is not able to deduct liquidated damages.[6]

In the *Gaymark* case, Gaymark as employers were responsible for a period of delay. The contract provided that the contractor had an entitlement to an extension of time for employer-caused delays but only as long as the contractor gave notice as required by the notice provisions in the contract. The contractor had failed to serve notice within the 14-day period required and Gaymark deducted liquidated damages because of the delay. The arbitrator held that the delays caused by the employer were acts of prevention which put the time for completion 'at large' and accordingly the employer had lost the right to deduct liquidated damages. On appeal the judge said:

> *Acceptance of Gaymark's submissions would result in an entirely unmeritorious award of liquidated damages for delays of its own making (and this in addition to the avoidance of [the contractor's] delay costs because of the company's failure to comply with the notice provisions.*

With this principle in mind, contractors might like to consider the option of not complying with notice provisions concerning a delay caused by an act of prevention by the employer, where such notice provisions are conditions precedent to a time extension being granted, and instead let time run at large, allowing them to complete within a reasonable time frame without the application of liquidated damages.

It should be noted that where the delay was not caused by an act of prevention by the employer the previous ruling in *Turner* v *Austotel*[7] would apply:

> *If a Builder, having a right to claim an extension of time fails to do so, it cannot claim that the act of prevention which would have entitled it to an extension of time for Practical Completion resulted in its inability to complete by that time. A party to a contract cannot rely upon preventing conduct of the other party where it failed to exercise a contractual right which would have negated the effect of that preventing conduct.*

In a similar case more recently in Scotland,[8] the Inner House of the Court of Sessions pointed out on appeal that the contractor in this case, who had similarly failed to issue a notice under an extension of time clause, was not in breach of contract in failing to issue a notice of delay arising out of an employer's instruction. The extension of time clause in the contract gave the contractor an additional right (the right to issue a notice within a specified time claiming its right to an extension of time) but did not oblige the contractor to invoke the clause's protection.

If not complied with, the clause had the effect of preserving the employer's right to liquidated damages for delay that would otherwise have been lost by an extension of time and the liquidated damages therefore remained payable by the contractor. They could not be considered as a penalty, as claimed by the contractor, simply because the contractor had failed to take the steps available to it. This was a case under Scottish law that went some way to contradicting the *Gaymark* decision in Australia.

Few standard forms of contract make giving a notice a condition precedent to granting an extension of time because of the danger that failure to comply with the notice clause could put time at large. One of the few standard forms that does make giving a notice a condition precedent is the KCRC[9] Conditions of Contract in Hong Kong and the implications of the *Gaymark* ruling may well cause a rethink of this when the next edition of the KCRC Conditions is under review.

1. *Banque Quilpue* v *Bryant* (1904).
2. *Peak Construction (Liverpool)* v *McKinney Foundations Ltd* (1970) 1 BLR 111, CA.
3. *Wells* v *Army & Navy Co-operative Society Ltd* (1902) 86 LT 764.
4. *Amalgamated Building Contractors Ltd* v *Waltham Holy Cross UDC* (1952) 2 All ER 452, CA at p.455.
5. *Fisher* v *Ford* (1840).
6. *Gaymark Investments Pty Ltd* v *Walter Construction Group Ltd* (1999) 16 CL 449; NTSC 143.
7. *Turner Corporation Ltd (Receiver and Manager appointed)* v *Austotel Pty. Ltd* (1994) 13 BCL 378.
8. *City Inn* v *Shepherd Construction Ltd* 2003 Scots CS 146.
9. Kowloon Canton Railway Corporation Conditions of Contract.

Time for Review

One of the problems which every contractor, where ever it is working in the world, will complain about is the time taken by the employer or its representative to review documentation, including drawings, method statements, change order estimates and so on and the damaging effect that this can have on its ability to maintain its programme. This is a truly international problem.

It frequently happens that the employer or its representative is either too busy to review the documentation submitted or is not sufficiently advanced with its own works or with the project funding. It then uses the delay in reviewing documents as a means of controlling progress. It is not unusual to see a contractor using this as a basis for a claim for extension of time but unless there is a specific contract clause that it can quote as empowering it in this way, and unless its submissions are totally without fault as to both content and timing, then it stands little chance of success.

On the other hand, some contractors make submissions too far in advance, sometimes when the employer's team is not sufficiently staffed or prepared to undertake their review, in an attempt to force the issue and make a case for delay on the part of the employer or its representatives. On other occasions the employer is hampered by incomplete submissions, submissions made out of sequence and submissions riddled with errors requiring several attempts to get them right, causing the employer additional expenditure of time and resources which had not been foreseen when entering into the contract.

The 'approval of documents' provisions in the contract usually contain a requirement that submissions should be made at a certain time, or at least within a reasonable period of time in relation to the date the work to which they relate has been programmed to commence, and that they should be responded to within a further stipulated period. If the submission is made unreasonably early in relation to the contract programme for the works, the employer should either return the submission as premature or give notice to the contractor that it will be reviewed but that the period for review will be in excess of that stipulated in the contract as a result of its premature submission.

The employer's responses must draw attention to any inaccuracies or deficiencies in the submissions but must at all times be reasonable and must take due account of the damaging effects a rejection or a qualified or delayed approval can have on progress. The contractor must, upon receipt of the employer's response, check its validity and either act upon it or, if it considers that it has adequate grounds for doing so, dispute it stating the reasons for doing so and its intention of claiming delay as a result.

The periods for review and approval can be tailored to meet the particular requirements of the contract and the project works, and we must at all times ensure that whatever is agreed as a workable procedural solution is adhered to strictly by all parties. In order to avoid subsequent disputes regarding delayed submissions or approvals, we should, at the beginning of the contract and with the support of and advice from the design team, planners, QA/QC and so on, establish a system for the submission of documents for approval.

This should be based upon a realistic submission schedule and should incorporate a monitoring system which will identify delays in review comments, approvals, rejections etc. in order that these can be recorded and notices served as soon as either party is observed as delaying the smooth running of the project, deliberately or not, or otherwise being unreasonable with its approach to approvals, including being over-pedantic in its interpretation of the requirements of the contract, over and above the intent of the contract.

It is helpful to include a clause in the contract that establishes clearly the principles to be adopted by both parties as regards approvals, which shall include:

- reference to a schedule, probably contained in the technical specification or the Appendix, that lists the documents that need to be submitted to the employer and the dates or time from commencement by which submissions have to be made;
- a number of working days within which the employer must either return them endorsed with its approval or notify the contractor in writing of the reasons for their disapproval, together with the amendments required;
- where the submissions were submitted unreasonably early with regard to the programme and the schedule of submissions mentioned above, a date by which the employer intends to respond if this is going to extend beyond the normal period noted above;
- a number of working days following the employer's return of the submitted documents within which the contractor is required to submit the amendments requested by the employer;
- a number of working days within which the employer must return the amended documents endorsed with its approval;
- provision for deemed approval where the employer fails to respond within the number of days set down for its response either at the initial or subsequent submission stage;
- provision that the employer will not disapprove any documents save only on the grounds that they do not comply with the requirements of the contract;
- provision that any material amendments requested to approved documents, or further material amendments beyond those already requested at the initial submission stage, will be covered by a change order.

T

Time of the Essence

The phrase 'time is of the essence' is often used in a commercial contract context; moreover, it is as often as not misused. A typical situation where time would be of the essence would be where a failure to complete by the contract completion date was a fundamental breach of the contract giving the employer the right to treat the contract as repudiated. The employer would in that case be able to terminate the contract and seek damages from the contractor resulting from his failure to complete.

This situation would not be a normal construction contract situation but could be envisaged where the particular structure to be completed by the contractor was intended to have a very specific use on, or commencing on, a very specific day. A good example would be the Olympic Games where new facilities would have to be completed in order that the Games could commence on the specific day set aside for their use and which had been the subject of many years' planning and organisation.

Time may be made of the essence by an express term of the contract or by implication. An express provision may be made by the inclusion of the words 'time is of the essence' in the agreement or by stating that failure of one party to perform within a stipulated time will entitle the other party to terminate the contract. In a contract for the hire of computers time was of the essence for the prompt payment of instalments. The Court of Appeal held that the failure to make one payment promptly was sufficient to evince repudiation of the contract entitling the hire company to terminate the contract and to recover in that case not only the late payment but also the loss of future instalments.[1]

Under the Sale of Goods Act 1979 in England and Wales time would not be of the essence in a contract for the sale of goods unless it was expressly stated as being so as in the case quoted above. Even so, if the express 'time of the essence' stipulation is deemed to be unfair and therefore unenforceable it can be overruled according to the Unfair Terms in Consumer Contracts Regulations (1999) in England and Wales.

If time is not intended to be of the essence it would be wise to insert an express provision to that effect, otherwise such a term might be implied in equity as shown below.

Time may be implied to be of the essence, either by the actions of the parties or by the terms of the contract. The two principal situations where this applies in equity[2] are:

- where the circumstances suggest that a completion date or period of time for completion must be complied with[3] and that failure to do so will go to the root of the contract and deprive the claimant of a substantial benefit of the contract which it was intended that it should receive;[4]
- where reasonable notice is given by one party that the other party has to complete by a certain date or within a certain period of time. This could occur where, for instance, a party failed to complete by the contract completion date. The other party could in that case serve a notice giving a reasonable further time for

completion, stating that if that further period is not met then the contract will be treated as repudiated.[5]

Most construction contracts, however, address the question of time for completion by the inclusion of clauses that facilitate the granting of extensions of time under certain circumstances and for payment of liquidated damages in others. This would seem to imply that the employer was perfectly content with the situation whereby excusable delays were remedied by extending the date for completion and whereby non-excusable delays resulted in it receiving liquidated damages as compensation for the delay.

Even where the contract states that time is of the essence the courts have been reluctant to accept that this actually goes to the essence of the contract.[6] Where the contractor's delay is major and where it has little idea as to when completion might be achieved, this might give rise to a situation where the employer was obliged to make time of the essence by serving a notice on the contractor rendering time of the essence.[7] There would, however, have to be strong and irrefutable evidence sufficient to evince that the contractor had no intention to be bound by the contract completion date before the courts would accept the employer's intention of making time of the essence. It would in any case not be acceptable to terminate the contract based only on the contractor's conduct without first giving the contractor due notice of the intention to do so.[8]

In a case where two parties owned two companies jointly, the parties agreed to each buy one of the companies outright and it was a condition precedent to completion that share certificates would be exchanged. Just before completion one party informed the other that it was not in possession of the share certificates in respect of the company that the other party was purchasing and asked the other party to check the statutory books for them. The other party did not respond until immediately prior to completion, when it stated that if the share certificates were not produced it would terminate the agreement and sue for damages.

The court held that although time was in all probability of the essence of the contract the party seeking to terminate had failed to cooperate with the other party and as such had been responsible for preventing the other party being in a position to complete. The party seeking to terminate lost its case, its right to terminate for the breach of the implied 'time of the essence' provision having been irrefutably damaged by it sharing blame for the delay.[9]

1. *Lombard North Central Plc* v *Butterworth* (1987) QB 527.
2. 'Equity' is defined in *Oran's Dictionary of the Law* as 'A system of courts that originated in England to take care of legal problems when the existing laws did not cover some situations in which a person's rights were violated by another person; a court's power to do justice where specific laws do not cover the situation'. Equity was originally administered by the Court of Chancery as a supplement to the rules and procedures of Common law.

Subsequently the Court of Chancery amalgamated with the Common Law Courts to form the Supreme Court by the enactment of the Judicature Act of 1873. Rules of equity are now administered in all divisions of the court in England and Wales. Rules of equity prevail over rules of law where there is a conflict.

3. *United Scientific Holdings* v *Burnley Borough Council* (1978) AC 904.
4. *Olympia & York Canary Wharf (No.2)* (1993) BCC 159.
5. *British & Commonwealth Group Plc* v *Quadrax Holdings Inc.* (1983) 3 All ER 492.
6. *Lucas* v *Godwin* (1837) 3 Bing.NC 737 – held that time being stated as of the essence did not mean there would be no payment if completion was not achieved in the time stated.
7. *Stickney* v *Keeble* (1915) AC 386, HL. One party guilty of undue delay was notified by the other party that unless performance was completed within a reasonable time the contract would be regarded as at an end.
8. *Felton* v *Wharrie* (1905).
9. *Grant* v *Cigman* (1996) 2 BCLC.

Without Prejudice

The intention behind the 'privilege' of pleading 'Without Prejudice' is to encourage settlements between parties in dispute. During the negotiation stages parties do not want to be inhibited by thoughts that every word mouthed, put on paper or otherwise communicated in an attempt to settle their dispute is somehow cast in tablets of stone. If this were to be the case the parties would be unlikely to enter into meaningful and open discussions on ways of resolving the dispute.

A 'without prejudice' communication need only be one step on the road towards resolving the dispute between the parties but it must be made with a view to settlement. The court will decide what qualifies as a communication 'with a view to settlement', taking into account the character of the individual communication and the context in which it is used, particularly in relation to other communications between the parties. The communication need not be so specific that it actually proposes a settlement figure but it must contribute significantly to the processes underway to reach such a figure.

It is becoming a habit in the industry to use the expression 'without prejudice' in attempts to prevent the recipient using the letter against the initiator should the matter concerned be, or later become, a dispute between the parties and where such communication might be seen as an admission of liability in later court proceedings. This privilege is not to be abused and is only to be used where the communication concerned is initiated with the definite intention of it being used and viewed as a genuine and constructive attempt at aiding the settlement of the dispute between the parties. Where there is such a genuine intention there is in fact not even the need to add the expression 'without prejudice'.

In general, a letter which contains an offer, responds to an offer or is written during genuine negotiations will have the protection of 'without prejudice' status even if the actual words are not imprinted upon it. However, in international contracts in particular, where not all of the parties are native English speakers, it is wise to add the words 'without prejudice' on relevant communications so as to save any misunderstandings at a later date.

The converse of the above is that communications which are issued when no dispute exists, which merely state one party's alleged rights and which are not part of an on-going negotiation situation, will not enjoy 'without prejudice' protection even if the words 'without prejudice' are imprinted on them.

From a strategic viewpoint, it might be preferable for a party to make 'open' communications (i.e. not stating 'without prejudice') since if the dispute reached court the open communications might well then be seen as evincing the genuine intention of the initiating party to reach a reasonable solution, i.e. not hiding behind 'without prejudice' offers that might never again be made.

Where 'without prejudice' protection is deemed to be beneficial or necessary, rather than merely stamp the communication concerned with the basic words it is suggested that a final paragraph in terms of the following would have more effect. However, it is advisable to take legal advice in the country of application to ensure that its use in the manner intended conforms to the legal requirements and interpretations then current in the country of application:

> *This letter is an effort by [Party 'A'] to compromise an ongoing dispute, and is subject to all restrictions and limitations on the use of such communications. The statements made and positions taken in this letter are without prejudice to positions that [Party 'A'] may take in the course of further negotiations, or in the course of any legal proceeding. [Party 'A'] hereby reserves all of its rights and remedies under the Contract, at law or otherwise, including its right to initiate an arbitration pursuant to Article [....] of the Contract.*

Appendices

Glossary of Terms

The intention of this Glossary of Terms is to explain some of the terms used both in this book and in the administration of international contracts in a form that can be utilised by either the employer or the contractor in the 'Definitions' section of contract and sub-contract documents. Their use as definitions would help in reducing the risk of doubt or confusion arising during the currency of the contract as to what is the true intention of the parties to the contract. It is appreciated that different types of contracts and projects will require different definitions; however, it is hoped that this Appendix will guide the reader towards the most suitable definition for a particular situation.

In multi-national contract situations where different nationalities are contracting together there cannot be enough clarification of the intent of the parties before the contract is entered into. Accurate definitions protect the employer from nebulous claims based on ambiguous or unclear wording of clauses, lack of clear definition or conflicts between different definitions, and similarly protect the contractor from erroneously acting upon an assumed interpretation of a term only to find later that its interpretation was incorrect.

This Glossary is not written in tablets of stone and the definitions given should be amended to suit the actual intentions of the parties in each particular project situation. Many of the words used in these definitions may well be unfamiliar to readers who do not have significant international experience and it is hoped that such words, phrases and intentions will be understood more clearly as a result of this Glossary.

Those familiar with writing contracts will have their own preferred ways of writing some of the definitions included in this Appendix and it is not my intention to suggest that such proven definitions are in any way deficient. If they have worked in the past, by all means continue using them as long as they still reflect accurately the current situation within the industry and the particular requirements of the project to which they refer.

It is left to the individual to judge from experience and from knowledge of the relevant country and of the parties concerned which of these definitions need to be referred to legal experts for confirmation prior to their incorporation into a contract. It is recommended that expert legal guidance should be sought whenever you are anything other than 100 per cent confident that the definition you are using is the right one for that particular application.

Accepted

When applied by Employer or Employer's Representative to Contractor's Drawings, Documents etc. (collectively referred to as 'documents'), means that the Employer or the Employer's Representative has not identified any statement or feature in or appertaining to the documents submitted by the Contractor which indicates that any aspect of them deviates from the requirements of the Contract.

Accepted as Noted

When applied by the Employer or Employer's Representative to Contractor's Drawings, Documents etc. (collectively referred to as 'documents'), means that, except as noted in the Employer's or Employer's Representative's comments, the Employer or Employer's Representative has not identified any statement or feature in or appertaining to the documents submitted by the Contractor which indicates, provided the comments noted by the Employer or Employer's Representative are incorporated by the Contractor into its documents, that any aspect of them deviates from the requirements of the Contract.

Achieved Milestone

The Milestone will be an Achieved Milestone as and from the date on which the Employer's Representative issues a certificate to both the Employer and the Contractor to the effect that it is satisfied, on the basis of its own inspections and information, that all matters which need to be carried out to satisfy a Milestone have been carried out.

Affiliate

Means, with respect to any person, any other person who alone or together with any other person directly or indirectly through one or more intermediaries controls, or is controlled by or is under common control with such person. For the purpose of this definition, 'control' (including the terms 'controlling', 'controlled by' and 'under common control with') as used with respect to any person shall mean the possession, directly or indirectly, of the power to direct or cause the direction of the management or policies of such person, whether through the ownership of voting securities, by contract, agency or otherwise.

Agent Bank

Means the person notified in writing from time to time to the Contractor by the Employer's Representative as the person appointed by the Lenders or any of them to act as agent or trustee, and who acts on behalf of the Lenders.

Appendix 1: Glossary of Terms

Applicable Laws and Regulations[1]

Means all published laws, treaties, ordinances, judgments, decrees, injunctions, writs and other orders of any governmental authority, and rules, regulations, orders or interpretations of any governmental authority having jurisdiction over the Parties, the Project or any activity conducted at or in connection with the Project or the Site, including use or disposal of any hazardous material, transportation of all Equipment and other items necessary for the Project to and from the Site, performance of the Services, start-up, testing of the Plant, as may be applicable and in effect from time to time.

Authority

Includes any:

- governmental, semi-governmental, local government or statutory authority;
- administrative or judicial body;
- tribunal, department, commission, public authority, agency, Minister, statutory corporation or instrumentality.

Award Date

Means the date on which the formal Contract, to which these General Conditions are attached, has been completed and signed by the Employer and the Contractor.

Business Day

Means a Day (other than a Saturday or a Sunday or a public holiday) upon which banks in the Country are open for business and, in respect of a Notice being served in a different country, a Day (other than a Saturday or a Sunday or a public holiday) upon which banks in that country are open for business.

Certificate of Final Completion

Means the certificate issued by the Employer pursuant to Clause [xxxx] certifying that the Work has met the requirements for Final Completion.

Change in Law

Means:

(i) the enactment or issuance of any new Law or Permit;
(ii) the amendment, alteration, modification or repeal of any existing Law or Permit;
(iii) any authoritative interpretation of any existing Law or Permit issued

by a competent court, tribunal, Government Authority or statutory authority contrary to its existing official interpretation;

in any case, with respect to (i), (ii) or (iii) above, coming into effect after the date of this Contract and which can reasonably be expected to have a material adverse effect on the obligations of the Parties under this Contract and that must be complied with in order for the Work to be constructed or for the Facility to be operated lawfully; provided, however, that any change in any national, state or other law relating to income or income withholding taxes shall not constitute a Change in Law for the purposes hereof.

Change in the Works

Means any omission from, addition to or other change, modification whatsoever to the Contract terms and conditions, the Specification or the Work (including any substitution or alteration and any change in quality, form, character, kind, position, location, dimension, level or line and any change of sequence, method or timing of any operation), which is instructed or approved in writing by the Employer under the Contract and which affects the Contract Price, the scheduled Provisional Acceptance Date, and any of the performance guarantees or any material right or obligation of either Party.

Change Order

Means a written order to the Contractor signed by the Employer authorising a Change in the Works.

Claim

Means any claim for an increase in the Contract Price, for payment of money (including damages) or for an extension of time:

- under, arising out of, or in any way in connection with, the Contract, including any direction of the Employer's Representative; or
- arising out of, or in any way in connection with, the Works or either Party's conduct before the Contract; or
- otherwise at law or in equity including:

  (i) by statute;
  (ii) in tort for negligence, including negligent misrepresentation; or
  (iii) for restitution.

Competent Authority

Includes any court of competent jurisdiction and any local or national or supranational agency, authority, department, government, legislature, minister, ministry, official or public or statutory person (whether autonomous or not) of any state or supranational organisation, including the European Union.

Completion

The stage when in respect of the Works:

(a) the Works are complete except for minor Defects:

- which do not prevent the Works from being reasonably capable of being used for their intended purpose;
- in relation to which the Employer's Representative determines the Contractor has reasonable grounds for not promptly correcting them;
- which can be corrected without prejudicing the convenient or lawful use of the Works;

(b) the Contractor has provided the Employer's Representative with all Approvals which it required to obtain under Clause [xxxx];
(c) the Employer's Representative has issued a Completion Certificate in accordance with Clause [xxxx]; and
(d) the Contractor has done everything which the Contract requires it to do as a condition precedent to Completion.

Consents and Approvals

Means all authorisations, consents, decrees, notices, directions, licences, visas, permits, waivers, privileges, approvals and the like from, and filings with, any Governmental Authority as may be required for, or in connection with the performance of the Contractor's obligations under this Contract, irrespective of whether all or any of them are required to be in the name of the Employer or the Contractor.

Contract Price

Means the sum stated in Sub-clause [xxxx] to be the Contract Price, as it may be adjusted in accordance with the Contract.

Contractor

Means [xxxx] and its legal successors in title, but not (except with the consent of the Employer) any assignee.

Contractor's Equipment

Means all machinery, apparatus, vehicles and other things required for the design, execution and other performance and provision of the Work, including the remedying of Defects, but does not include 'Materials, Plant or Equipment' or 'Temporary Work'.

Contractor Good Industry Practice

Means the exercise of that degree of skill, diligence, prudence and inspection and maintenance practice which would reasonably be expected in the Country from a skilled and experienced Contractor providing services of the same type as the services provided by the Contractor in accordance with the terms of this Contract under the same or similar circumstances.

Contractor's Information

Means all documents, reports, schedules or information (and all revisions thereof) which the Contractor is required under the Contract to submit to the Employer's Representative or to the Employer.

Contractor's Key Personnel

Means those personnel designated by Contractor and approved by Employer who shall responsible for managing day to day activities for the execution of the Works.

Contractor's Representative

Means the person approved in writing by the Employer's Representative and notified by the Contractor to the Employer's Representative from time to time as the person appointed by the Contractor to act as the Contractor's Representative for the purposes of the Contract.

Cost

Means all expenditure properly incurred (or to be incurred) by the Contractor whether on or off the Site, including overhead and similar charges, but does not include profit.

Country

Means the country in which the Works are to be executed and to which the Plant and Materials are to be delivered.

Date for Completion

Means in respect of the Works, the date for completion specified in the Contract Particulars, as adjusted under the Contract.

Date of Completion

Means the date of Completion as set out in the Notice of Completion.

Day

Means a period of 24 hours commencing at 00.00 hours, and ending at 23.59 hours of one calendar day.

Default

Means the failure to observe a provision of this Contract.

Defect or Deficiency

Unless otherwise specifically stated herein:

- 'Defect' means any design, engineering, Equipment, supply, test, procedure or assembly which does not conform to the Work Scope, Employer's Standards or the Drawings and Specifications, including any improper or inferior workmanship that is inconsistent with the Work Scope, the Drawings and Specifications or the Employer's Standards;
- 'Deficiency' refers to any shortfall in performance outside the permitted tolerances of the performance guarantees during the performance tests carried out in accordance with the procedures agreed to between the Employer and the Contractor, consistent with the standards and operating conditions for performance testing set forth in the Contract, or during the performance tests any other condition that is under the reasonable control of the Contractor that would materially and adversely affect either the safe and continuous operation of the Equipment during the Equipment's design life or, as warranted, the performance of the Equipment under normal operating conditions.

Defects Liability Period

Means the period from the Provisional Acceptance Date until [xx] Months after the Provisional Acceptance Date, as may be extended under Clause [xxxx].

Delivery Schedule

Means the dates set forth in the Contract for the delivery of the components of the Material, Plant and Equipment.

Direction

Means any decision, demand, determination, direction, instruction, notice, order, rejection or requirement.

Dissolution

Dissolution of a person includes the bankruptcy, insolvency, liquidation, winding-up, administration or dissolution of that person and the appointment in respect of that person or any of its assets of a receiver, administrator or similar officer and any proceeding or event which is equivalent or analogous to any of the same by whatever name known and in whatever jurisdiction and any step taken for or with a view to any of the foregoing.

Dispute

Means any dispute, difference or claim arising out of or connected with the Contract or the Work (including as to any decision, determination or instruction of the Employer's Representative and any as to the validity of the Contract or any provision thereof) and whether before or after the Final Completion date or termination of the Contract or the Contractor's engagement there under.

Documents

Means all written or graphic matter or other means of preserving thought or expression, including without limitation, books, papers, files, notes, correspondence, memoranda, drawings, plans, charts, procedures, certificates, specifications, reports, studies, summaries, written information, data, computer printouts, microfilm, microfiche, audio tapes, video tapes, magnetic tapes, computer diskettes and disks and other electronic media and computer-stored information.

Drawings and Specifications

Means drawings and specifications either:

- described in or attached to this Contract; or
- prepared by the Contractor or any of its Sub-Contractors or Suppliers with respect to the Work.

Effective Date

Means the date on which the Contract entered into legal force.

Employer

Means Employer, Employer's Representative, Lenders, Lender's Engineer and all affiliates, officers, directors, employees and agents thereof.

Employer Caused Delay

Means any delay or cost impact in the performance and completion of the Work that is directly caused by an act, omission or failure by Employer to perform its obligations hereunder, including the removal of Hazardous Materials on the Site and delay in obtaining Permits which the Employer is required to obtain hereunder, provided, however, that Employer Caused Delay shall not include:

(a) any act or omission that is permitted under, and is taken or caused in accordance with, the terms of this Contract;
(b) any act or omission of Employer or its agent acting under or in accordance with Contractor's written instructions; or
(c) any delay in the Employer's performance of an obligation to be performed by a specified date hereunder if a Change Order has been issued delaying a date by which Contractor is required to perform an event hereunder (provided the Employer shall not be excused for a period longer than the period that the Contractor is excused).

Employer's Representative

Means the person notified in writing by the Employer to the Contractor from time to time as the person appointed by the Employer to act as the Employer's Representative for the purposes of the Contract.

Expert

Means an independent expert engaged for the purpose of resolving a dispute under the Contract.

Extra Cost

Means additional direct expenditure reasonably and properly incurred by the Contractor plus an allowance of up to [ten] per cent thereof for overheads reasonably and properly allocatable thereto, but does not include profit nor any expenditure or overhead covered by insurance proceeds received or to be received.

Facility

Means the whole of the Work to be executed and supplied by the Contractor in accordance with this Contract, including Changes and all Materials, Plant and Equipment which is to be handed over to the Employer in accordance with the requirements of this Contract.

Final Completion

Final Completion is defined in Clause [xxxx].[2]

Final Completion Certificate

Final Completion Certificate is defined in Clause [xxxx].[3]

Final Completion Date

Means the Day stated in the Final Completion Certificate as the Day on which the Work achieved Final Completion or, in the case of deemed Final

Completion, the Day on which the Work is deemed to have achieved Final Completion.

Financial Close

Means the date when the facilities under the Loan Agreement are available for use by the Employer and the conditions precedent to first draw-down have been satisfied or waived.

Finance Agreements

Means the documents and arrangements relating to the provision of credit, finance and other financial accommodation or facilities to the Employer in connection with the development, operation and/or maintenance of the Facility or the provision of security for such credit, finance and other financial accommodation or facilities.

Foreign Currency

Means a freely convertible currency, named in the Contract as a currency in which part of the Contract Price is payable, but not the Local Currency.

Good Operating Practices

Means the practices of a person seeking in good faith to perform its contractual obligations and exercising that degree of skill, diligence, prudence and foresight which would reasonably and ordinarily be expected from a skilled and experienced operator complying with the legal requirements engaged in the same type of undertaking in the same or similar circumstances or conditions.

Government Authority

Means the government of the Country and any ministry, department or political subdivision thereof, and any person exercising executive, legislative, judicial, regulatory or administrative functions of or pertaining to government or any other governmental entity, instrumentality, agency, authority, corporation, committee or commission under the direct or indirect control of the government of the Country.

Government Approvals

Means authorisations, consents decrees, notices, directions, licences, visas, permits, waivers, privileges, approvals and the like from, and filings with, any Governmental Authority, including those listed in the Specification.

Hazardous Materials

Means any hazardous materials including any form of organic or chemical matter whether solid, liquid, gas, odor, heat, sand vibration, radiation, or substance which makes or has the capacity to make the environment:

(a) unsafe or unfit for habitation or occupation by persons or animals;
(b) degraded in its capacity to support plant life;
(c) contaminated; or
(d) otherwise environmentally degraded

or which is defined or included in the definition of 'hazardous substance', 'hazardous waste', 'hazardous material', 'extremely hazardous waste', 'restricted hazardous waste', 'toxic substance' or 'toxic pollutants' or any combination of such or similar words under any applicable environmental law.

Import Duties

Means all duties, charges, levies, taxes and imposts imposed in respect of the importing of any goods or materials into the Country for, or in connection with, the Work.

Information

Includes know-how, knowledge, data and information whatsoever and howsoever held.

Insolvency

Means any one of the following:

(a) notice is given of a meeting of creditors with a view to the Contractor entering into a deed of company arrangement;
(b) the Contractor entering a deed of company arrangement with creditors;
(c) a controller or administrator is appointed to the Contractor;
(d) an application is made to a court for the winding up of the Contractor and not stayed within 14 days;
(e) a winding up order is made in respect of the Contractor;
(f) the Contractor resolves by special resolution that it be wound up voluntarily (other than for member's voluntary winding-up); or
(g) a mortgagee of any property of the Contractor takes possession of that property.

Intellectual Property Rights

Means:

(a) copyright,[4] patent, registered design, design right, trade mark, trade name, trade secret or other industrial or intellectual property right; and

(b) any right to bring an action for passing off or any similar or analogous proceeding.

Interested Parties

Means the Employer, the Employer's Representative, Employer's technical advisers, the Lenders, the Lenders' Engineer, the Agent Bank and their respective agents and employees.

Job Site

Means that real property on which the Facility is to be located, plus any transmission or pipelines or other rights-of-way and such additional areas as may, from time to time, be designated in writing by Employer and/or Employer's Representative for Contractor's use

Late Payment Rate

Means in respect of foreign currency, the rate of [two] per cent per annum above LIBOR[5] for Foreign Currency as published by the Financial Times in the UK on the respective day and, in the case of local currency payments, [two] per cent above the official discount rate of the Country's Central Bank on the due day.

Law

Means:

(a) any applicable statute, law, rule, regulation, code, ordinance, judgment, decree, writ, order or the like of any Government Authority and the official interpretations thereof; or

(b) any official requirements or conditions on or with respect to the issuance, maintenance or renewal of any applicable Permits or licenses or applications there for maintained by any Government Authority.

Lenders

Means the persons providing credit, finance or other financial accommodation or facilities, or receiving security, under the Financing Agreements, and includes the Agent Bank.

Lender's Engineer[6]

Means the person notified from time to time to the Contractor by the Employer's Representative as the person appointed by the Lenders to act as the Lenders' Engineer for the purposes of the Contract.

Liability

Includes any cost, expense, damage, injury, loss, claim, action, proceeding or liability whatsoever.

Local Currency

Means the lawful currency of the Country, at the date of the Contract.

Major Sub-Contract

Means any sub-contract (or group of sub-contracts relating to the same matter) with a single sub-contractor with a value or cost to the Contractor in excess of ten per cent of the Contract Sum or with a duration in excess of one year.

Material Change

Means any proposed Change Order or other amendment, supplement or modification to this Contract which, directly or indirectly:

(a) delays or extends any Milestone, the performance test completion deadline or the Final Completion Date;
(b) modifies the performance guarantees or any component thereof or the manner in which such performance is tested or measured;
(c) alters the Milestone Payment Schedule;
(d) alters the scope or duration of any Contractor warranty;
(e) alters the amount, timing or manner of payment of any Contractor liquidated damages;
(f) causes any portion of the Facility to materially deviate from the scope of the Work;
(g) increases the Contract Price by more than [xxx] for an individual change or, in conjunction with all other changes previously effected, causes the Contract Price to increase by more than [xxx] over the original Contract Price; or
(h) otherwise materially alters or waives any liability or obligation of the Contractor under this Contract or any right or benefit of Contractor and/or Employer hereunder.

Materials, Plant and Equipment

Means all materials, plant, equipment, supplies, parts, apparatus, electrical and mechanical equipment, machinery, articles and things of all kinds to be provided and incorporated into the Works by the Contractor under the Contract but does not include Construction Equipment, special testing apparatus and other items that do not form part of the Permanent Work installation.

Milestone

Means an event listed in the Milestone Payment Schedule as a Milestone, which when accomplished will entitle the Contractor to apply in accordance with the Contract for the payment of the Milestone Payment shown in the Milestone Payment Schedule as being payable upon such accomplishment.

Milestone Payment

Means a sum listed in the Milestone Payment Schedule as a Milestone Payment.

Milestone Payment Schedule

Means the schedule set out in Schedule [xxx], as it may be amended in accordance with the Contract.

Month

Means a calendar month; the time of commencement of any calendar month period shall be at 00.00 hours on the applicable day.

Notices

Means all notices and other communications which, under the Contract, are required to be in writing and all other documents prepared or arising under or in connection with the Contract, including the Contractor's Documents.

Notice of Completion

Means a notice under Clause [xxxx] issued by the Employer's Representative stating that Completion has been achieved.

Notice to Proceed

Means a Notice under Clause [xxxx] issued by the Employer's Representative allowing the Contractor to commence the performance of the Works.

Permanent Work

Means the permanent work to be designed, executed and otherwise performed and provided by the Contractor under the Contract.

Permit

Means any authorisation, consent, approval, licence, ruling, permit, exemption, filing, variance, order, judgment, decree, publication, condition, notice to, declaration or registration of or with or regulation by or of any

Government Authorities relating to the acquisition, ownership, occupation, construction, start-up, testing, operation or maintenance of the Facility.

Person

Includes any individual, corporation, firm, partnership, joint venture, association, organisation, trust, estate, Government Authority or statutory person (whether autonomous or not) of any state (in each case whether or not having a separate legal personality).

Provisional Acceptance

Provisional Acceptance is defined in Clause [xxxx].[7]

Provisional Acceptance Certificate

Provisional Acceptance Certificate is defined in Clause [xxxx].[8]

Provisional Acceptance Date[9]

Means the Day stated in the Provisional Acceptance Certificate as the Day on which the Work achieved Provisional Acceptance or, in the case of deemed Provisional Acceptance, the Day on which the Work is deemed to have achieved Provisional Acceptance under the terms and conditions of this Contract.

Punch List

Means the list jointly prepared by the Parties at Provisional Acceptance or within five Days thereafter (provided that if as of such date the Parties have not reached agreement on the inclusion or exclusion of any items, the Punch List shall be deemed to include all such disputed items as the Employer believes should be included [as noted by the Employer in writing to the Contractor and stating in reasonable detail its reasons therefore] unless and until it is otherwise agreed by the Parties or otherwise determined under the Contract, which list shall set forth all items of Work which remain to be performed in order to ensure that the Equipment fully complies with all of the standards and requirements set forth herein. The Punch List shall not include any items of Work, alone or in aggregate, the non-completion of which prevents the Equipment from being used legally and safely.

Punch List Items

Means items of Work identified by the Employer in writing on or prior to the Provisional Acceptance Date which have not been completed but which do not affect the safe start-up and testing or commercial operation of any component of the Facility and which are to be corrected, fixed or repaired by the Contractor at its own expense as a condition to achieving Final Completion.

Site

Means the places provided by the Employer where the Works are to be executed and to which Materials, Plant and Equipment are to be delivered, and any other places as may be specifically designated in the Contract as forming part of the Site.

Statutory Requirement

Means an Act of Parliament, regulation, statutory instrument and code and includes any licence, order, official policy, directive, request, requirement or guideline of the government or any Government Authority.

Sub-Contractor

Means a sub-contractor engaged by the Contractor to perform part of the Works and includes a supplier of goods or services (including professional services and plant hire).

Tax

Includes any present or future tax, levy, impost, duty, charge, fee, deduction or withholding of any jurisdiction, and any interest or penalty in respect thereof.

Temporary Work

Means all temporary works of every kind, including temporary buildings and structures, (other than Contractor's Equipment) required for the design, execution and other performance and provision of the Permanent Work and the remedying of Defects.

Work(s)

Means:

(a) the Permanent Work and the Temporary Work including without limitation, the provision of engineering, design (including design development and revision), procurement, fabrication, testing, shipping, storage, construction, erection, start-up, commissioning and testing services and associated management services and supply of equipment and training services and all other work, services and things, whether of a temporary or permanent nature (and including Materials, Plant and Equipment and Contractor's Documents) and the remedying of Defects, to be performed or provided by the Contractor under the Contract, but excluding the Contractor's Equipment; and

(b) all other work, services and obligations required to be carried out, performed or complied with by the Contractor as specified in the Specification.

Worker

Means any employee or agent of, or other person engaged by, the Contractor or any Sub-Contractor.

1. This definition is not to be confused with the 'Applicable Laws' section on pp.21 to 23 inclusive. This definition is a general one aimed at defining the meaning of the term when used in contexts such as 'Contractor shall comply with all applicable laws and regulations'. The 'Applicable Law' under which the contract shall be construed (choice of a law of a particular country) will still be the one adopted in the Applicable Law clause.
2. The definition of Final Completion will be defined within the body of the contract and will be compiled to take account of the particular circumstances of each contract.
3. The same applies to Final Completion Certificate.
4. A copyright is an unregistered right, whereas patents, registered designs or trade marks have to be registered. There is no official register for copyright and so there is no official action to take (i.e. no application to make, forms to fill in or fees to pay). Copyright comes into effect immediately, as soon as something that can be protected is created and 'fixed' in some way such as on paper, on film, via sound recording, as an electronic record on the internet, etc.

   It is a good idea to mark copyright work with the copyright symbol © followed by your name and the date, to warn others against copying it, but this is not legally necessary in the UK

   The works are protected by copyright regardless of the medium in which they exist and this includes the internet. Copyright does not protect ideas – copyright protects the way that ideas are expressed in a piece of work, but it does not protect the idea itself.
5. London Inter Bank Offered Rate – this can be changed as appropriate to whatever recognised Central Bank rate or similar is considered most suitable for a particular contractual relationship.
6. Also sometimes referred to as 'Independent Engineer'.
7. The definition of Provisional Acceptance will be defined within the body of the contract and will be compiled to take account of the particular circumstances of each contract. In some cases it may be called 'Taking Over', or some other title preferred by or common usage within a particular industry, profession or country.
8. The same applies to Provisional Acceptance Certificate which may be called 'Taking Over Certificate' or similar.
9. Similarly to the notes in the preceding footnotes, the definition title may sometimes refer to the 'Taking Over' Date or some other wording similar in meaning – references to 'Provisional Acceptance' and 'Provisional Acceptance Certificate' would need to be changed accordingly in this definition to match the actual definition title used.

International Legal Systems

Country:	Basis of Legal System:	Notes:	Parties to the New York Convention **
Algeria	Socialist Based on French and Islamic law	Republic; Judicial review of legislative acts in ad hoc Constitutional Council composed of various public officials including Supreme Court justices	Yes (Reservations 1,2)
Argentina	Mixture of US and Western European Legal systems	Republic	Yes (Reservations 1,2)
Armenia	Civil Law	Republic; Judicial review of legislative acts and constitutionality of laws by Constitutional Court; Court of Cassation is appeals court	Yes (Reservations 1,2)
Australia	English Common Law	Democratic Federal-state system of government; High Court – chief justice and six other justices appointed by Governor General	Yes
Austria	Civil Law influenced by Roman Law origins	Federal republic; Judicial review of legislative acts by Constitutional Court; Appeals from superior court judgments to one of four Courts of Appeal, further appeal possible to Supreme Court; Supremacy of European Community Law over Austrian national law is recognised.	Yes
Azerbaijan	Civil Law	Republic	Yes
Bahamas	English Common Law	Constitutional parliamentary democracy	No
Bangladesh	English Common Law	Parliamentary democracy	Yes
Belarus	Civil Law	Republic	Yes
Belgium	Civil Law, influenced by English constitutional theory	Federal parliamentary democracy under constitutional monarch; Judicial review of legislative acts	Yes (Reservation 1)
Bosnia & Herzegovina	Civil Law	Emerging federal democratic republic	Yes (Reservations 1,2)
Brazil	Roman Codes used as a basis	Federative Republic	Yes

Brunei	English Common Law & Islamic Law	Constitutional sultanate; Muslims are bound by Islamic Law in many circumstances	Yes (Reservation 1)
Cambodia	Civil Law; French Law influence coupled with royal decrees and acts of the legislature	Multiparty democracy under constitutional monarchy; Increasing influence of Common Law in recent years	Yes
Cameroon	French Civil Law with an element of Common Law influence	Unitary republic, multiparty presidential regime (weight of power remains with president)	Yes
Canada	English Common Law except in Quebec where Civil Law system based on French Law prevails	Confederation with parliamentary democracy	Yes
Cayman Islands	English Common Law & Statutes	British crown colony	No
Chad	French Civil Law & local customary law	Republic	No
Chile	Spanish Civil Law influenced by French and Austrian Civil Laws	Republic; Judicial review of legislative acts in the Supreme Court; new US-style adversarial style of criminal justice being introduced	Yes
China	Complex mix of custom, statute and Civil Law influence from Germany and Japan	Communist state	Yes (Reservations 1,2)
Congo – Republic of	French Civil Law & local customary law	Republic	No
Cote d'Ivoire	French Civil Law & local customary law	Republic, multiparty regime; Judicial review of legislative acts by Constitutional Chamber of the Supreme Court	Yes
Croatia	Civil Law	Developing presidential/parliamentary democracy	Yes (Reservations 1,2)
Czech Republic	Austro-Hungarian Civil Law	Parliamentary democracy; Legal code modified to bring in line with OSCE obligations and expunge Marxist-Leninist legal theory	Yes

Denmark	Civil Law	Constitutional monarchy; Judicial review of legislative acts	Yes (Reservations 1,2)
Djibouti	French Civil Law, Islamic Law and traditional practices	Republic	Yes
Dominica	English Common Law	Parliamentary democracy, republic within Commonwealth	Yes
Ecuador	Civil Law	Republic	Yes (Reservations 1,2)
Egypt	English Common Law, Islamic Law and French Napoleonic Codes	Republic; Judicial review by Supreme Court and Council of State (oversees validity of administrative decisions)	Yes
Estonia	Civil Law	Parliamentary republic; No judicial review of legislative acts	Yes
Finland	Swedish Civil Law	Republic; Supreme Court may request legislation interpreting or modifying laws	Yes
France	Civil Law with indigenous influences	Republic; Judicial review of administrative but not legislative acts	Yes (Reservation 1)
Gabon	French Civil Law & local Customary Law	Republic, multiparty presidential regime; Judicial review of legislative acts by Constitutional Chamber of the Supreme Court	No
Georgia	Civil Law	Republic	Yes
Germany	Civil Law with influences of German and Belgian Civil Codes and customary law	Federal republic; Judicial review of legislative acts by Federal Constitutional Court	Yes (Reservation 1)
Ghana	English Common Law & Customary Law	Constitutional democracy	Yes
Greece	Codified Roman Law	Parliamentary republic; Judicial divided into civil, criminal and administrative courts	Yes (Reservations 1,2)
Grenada	English Common Law	Constitutional monarchy with UK-style parliament	No
Guatemala	Civil Law	Constitutional democratic republic; Judicial review of legislative acts	Yes (Reservations 1,2)

Guinea	French Civil Law, local Customary Law and Decree	Republic; Legal codes under revision	Yes
Guyana	English Common Law with elements of Roman-Dutch Law	Republic within Commonwealth	No
Hong Kong	English Common Law	Special Administrative Region of China; Basic Law of Hong Kong adopted 4 April 1990, previous laws in force maintained in so far as they do not contravene Basic Law	No
Iceland	Danish Civil Law	Constitutional republic	Yes
India	English Common Law & Hindu Law	Federal republic; Judicial review of legislative acts in certain limited circumstances	Yes (Reservations 1,2)
Indonesia	Roman and Dutch Law substantially modified by indigenous concepts	Republic; New criminal procedures being introduced	Yes
Iran	Islamic Principles of Government codified by Constitution	Theocratic republic	Yes (Reservations 1,2)
Ireland	English Common Law modified considerably to accommodate local customs	Republic; Judicial review of legislative acts in Supreme Court	Yes (Reservation 1)
Israel	English Common Law modified in personal matters by Jewish, Christian and Muslim legal systems	Parliamentary democracy	Yes
Italy	Civil Law	Republic; Appeals are treated as new trials, judicial reviews a possibility in certain circumstances under Constitutional Court	Yes
Jamaica	English Common Law	Constitutional parliamentary democracy	Yes (Reservations 1,2)

Japan	European Civil Law influenced by Anglo-American Common Law; Customary Laws	Constitutional monarchy with parliamentary government; Judicial review of legislative acts by Supreme Court. Only Supreme Court decisions have effect on later interpretations of law. Constitution includes bill of rights similar to US Bill of Rights	Yes (Reservation 1)
Jordan	Islamic Law & French Civil Law	Constitutional monarchy; Judicial review of legislative acts in specially provided High Tribunal	Yes
Kazakhstan	Civil Law	Republic	Yes
Kenya	Based on Kenyan statutory and common law, Kenyan and English Common Law, tribal law and Islamic Law	Republic; Judicial review of legislative acts by High Court	Yes
Korea, North	German Civil Law modified by Japanese influences and Communist legal theories	Authoritarian socialist, one-man dictatorship; No judicial review of legislative acts	No
Korea, South	Combines continental European Civil Law systems with English/US law systems and Chinese classical thoughts	Republic	Yes (Reservations 1,2)
Kuwait	Civil Law, with Islamic Law predominant for personal matters; Commercial Law has French and Egyptian influences	Nominal constitutional monarchy	Yes (Reservation 1)
Kyrgyzstan	Civil Law	Republic	Yes
Laos	Socialist, influenced by traditional customs, French legal norms and procedures and Socialist practice	Communist state	Yes
Latvia	Civil Law	Parliamentary democracy	Yes

Lebanon	Ottoman Law, canon law, Napoleonic Codes, Civil Law	Republic; No judicial review of legislative acts	Yes (Reservation 1)
Lesotho	English Common Law & Roman-Dutch Law	Parliamentary constitutional monarchy; Judicial review of legislative acts in the High Court and Court of Appeal	Yes
Libya	Islamic Law and Italian Civil Law	Military dictatorship (in theory a 'Jamahiriya' – State of the masses governed by the populace through local councils); No constitutional provisions for judicial review of legislative acts	No
Lithuania	Civil Law	Parliamentary democracy; Review of legislative acts possible upon appeal to Constitutional Court	Yes
Luxembourg	Civil Law	Constitutional monarchy	Yes (Reservation 1)
Madagascar	French Civil Law & Malagasy Law	Republic	Yes (Reservations 1,2)
Malawi	English Common Law & Customary Law	Multiparty democracy; Judicial review of legislative acts in the Supreme Court of Appeal; Executive exerts considerable pressure over legislature	No
Malaysia	English Common Law	Constitutional monarchy; Judicial review of legislative acts in the Supreme Court upon request of supreme head of Federation	Yes (Reservations 1,2)
Mali	French Civil Law & local customary law	Republic; Judicial review of legislative acts by Constitutional Court	Yes
Malta	English Common Law & Roman Civil Law	Republic	Yes (Reservation 1)
Mauritius	French Civil Law with some elements of English Common Law	Parliamentary democracy	Yes (Reservation 1)
Mexico	Mixture of US constitutional theory and Civil Law	Federal republic; Judicial review of legislative acts	Yes

Morocco	Islamic Law and French/Spanish Civil Law	Constitutional monarchy; Judicial review of legislative acts in Constitutional Chamber of Supreme Court	Yes (Reservation 1)
Netherlands	Civil Law incorporating French penal theory	Constitutional monarchy; Judicial review forbidden by constitution	Yes (Reservation 1)
New Zealand	English Common Law	Parliamentary democracy; Special legislation and land courts for Maoris	Yes (Reservation 1)
Nicaragua	Civil Law	Republic; Judicial review of administrative (but not legislative acts) by Supreme Court	No
Niger	French Civil Law & local Customary Law	Republic	Yes
Nigeria	English Common Law & Traditional Law (Islamic law only applicable in some northern states)	Republic – transitional from military to civilian rule	Yes (Reservations 1,2)
Norway	Civil Law, Customary Law & Common Law	Constitutional monarchy; Supreme Court gives advisory opinions to legislature upon request	Yes (Reservation 1)
Oman	English Common Law & Islamic Law; French and Egyptian influences in commercial law	Monarchy; Ultimate appeal to Monarch	Yes
Pakistan	English Common Law modified to account for Pakistan being an Islamic state	Federal Republic	Yes
Panama	Civil Law	Constitutional democracy; Judicial review of legislative acts in the Supreme Court of Justice	Yes
Papua New Guinea	English Common Law	Constitutional monarchy with parliamentary democracy	No
Peru	Civil Law	Constitutional republic	Yes
Philippines	Spanish Civil Law and Anglo-US Laws	Republic	Yes (Reservations 1,2)

Poland	Mixture of continental Europe (Napoleonic) Civil Law and communist legal theories; Changes being introduced as part of democratisation process	Republic; Limited judicial review of legislative acts; Rulings of Constitutional Tribunal are final; Court decisions can be appealed to European Court of Justice in Strasbourg	Yes (Reservations 1,2)
Portugal	Civil Law	Parliamentary democracy; Constitutional Tribunal able to review constitutionality of legislation	Yes (Reservation 1)
Qatar	Discretionary system of law controlled by Emir; Islamic Law for personal and family matters	Traditional monarchy; Civil Codes are gradually being implemented	Yes
Russia	Civil Law	Federation; Judicial review of legislative acts	Yes
Samoa	English Common Law & Local Customs	Constitutional monarchy; Judicial review of legislative acts with respect to the fundamental rights of its citizens	No
Saudi Arabia	Islamic Law	Monarchy; Several secular codes have been introduced; Special committees handle commercial disputes	Yes (Reservation 1)
Senegal	French Civil Law	Republic under multiparty democratic rule; Judicial review of legislative acts in Constitutional Court	Yes
Serbia and Montenegro	Civil Law	Republic	No
Seychelles	English Common Law, French Civil Law and customary laws	Republic	No
Singapore	English Common Law	Parliamentary republic	Yes (Reservation 1)
Slovakia	Austro-Hungarian Civil Law	Parliamentary democracy; Legal code modified to bring in line with OSCE obligations and expunge Marxist-Leninist legal theory	Yes

Slovenia	Civil Law	Parliamentary democratic republic	Yes (Reservations 1,2)
South Africa	Roman-Dutch Law and English Common Law	Republic	Yes
Spain	Civil Law with regional influences and applications	Parliamentary monarchy	Yes
Sudan	English Common Law & Islamic Law	Authoritarian regime; Islamic Law applies to residents of Northern states regardless of their religion; Some special religious courts	No
Sweden	Civil Law influenced by customary law and German Law	Constitutional monarchy	Yes
Switzerland	Civil Law influenced by customary law	Federal republic; judicial reviews of legislative acts with exception of federal decrees of general obligatory character	Yes
Syria	Islamic Law and Civil Law	Republic with military regime; Some special religious courts	Yes
Taiwan	Civil Law strongly related to German Civil Law	Multiparty democratic regime	No
Tajikistan	Civil Law	Popular Democratic/Socialist and Sovereign State with Republican system of government; No judicial review of legislative acts	No
Tanzania	English Common Law	Republic; Judicial review of legislative acts in matters of interpretation only	Yes (Reservation 1)
Thailand	Civil Law with Common Law influences	Constitutional monarchy	Yes
Trinidad and Tobago	English Common Law	Parliamentary democracy; Judicial review of legislative acts in Supreme Court	Yes (Reservations 1,2)
Tunisia	French Civil Law & Islamic Law	Republic; Judicial review of legislative acts in certain circumstances by Supreme Court	Yes (Reservations 1,2)

Turkey	Derived from various continental European legal systems	Republican parliamentary democracy	Yes
Turkmenistan	Civil Law	Republic; Adversarial court system	No
Ukraine	Civil Law	Republic; Judicial review of legislative acts	Yes
United Arab Emirates	Federal Court system	Federation with specified powers delegated to the UAE federal government and other powers remaining with member emirates	No
United Kingdom	Common Law tradition with early Roman and modern continental influences	Constitutional monarchy; Judicial review of legislative acts	Yes (Reservation 1)
United States of America	English Common Law (except for Louisiana where Civil Law system prevails due to the state's earlier ties with France)	Constitution-based federal republic; Judicial review of legislative acts	Yes (Reservation 1)
Vietnam	Socialist based on communist legal theory and French Civil Law system	Communist State; Judicial review by Supreme People's Court	Yes (Reservations 1,2)
Yemen	Islamic Law, Turkish Law, English Common Law & Tribal Customary Law	Republic	No
Zambia	English Common Law & Customary Law	Republic; Judicial review of legislative acts in an ad hoc Constitutional Council	Yes

** **Notes in Column Relating to Parties to the Convention (New York Convention) on the Recognition and Enforcement of Foreign Arbitral Awards:**

Reservation 1: Awards will be recognised and enforced only if made in the territory of another Contracting State.

Reservation 2: Convention applies only to differences arising out of legal relationships, whether contractual or not, which are considered as commercial under the national law of the state making such declaration.

Appendix 3

Contractor's Pre-Tender Site Review[1]

1. Project Data

Job No:	Project Name:
Employer:	Project Location:
Employer's Representative:	Project Description:

2. Pre-Tender Site Inspection

Date of Visit: Start date:…………………..	Finish date:……………………..	Contractor Represented by:
Employer Represented by:		Others in Attendance:

3. Area Information[2]

3.1 Principal Cities Within 100km of Site:		
Name:	Population:	Distance from Site:

3.2 National and Local Government Status		
Political Stability:[3]	Support for Project-Nationally:[4]	Support for Project-Locally:[5]

3.3 Banking		
Local Banks in Vicinity:	Bank Charges:	Fund Transfer Facilities:
1.		
2.		
Foreign Banks in Vicinity:	Bank Charges:	Fund Transfer Facilities:
1.		
2.		

3.4 Culture		
Predominant Local Religions:	Nationality of Labour Available Locally:	Mix of Races Locally:

3.5 Public Relations	
Assess Local Feelings in Relation to Project:[6]	Assess media support for project:[7]

4. TAXES, CURRENCY, LICENCES, REGULATIONS ETC.

4.1 Taxes		
Corporate taxes:	Sales, business, turnover etc. taxes:	Income tax on company activities:
Income tax applied to local employees:	Income taxes applied to foreign specialists:	Exemptions:

Company's withholding tax responsibility:	Other taxes:	Other Taxes – Exemptions:

4.2 Currency		
Convertibility:	Exchange Restrictions:	Bank Rates: Sale: Purchase:
Open Market Rates: Sale: Purchase:	Contract Fixed Rate:	Contract Subsidized Rate:
Hard or dollar currency restrictions:	Exchange Rates Fluctuations Trends Over past 3–5 years:	Exchange Rate Risk During Project:

4.3 Licensing and Registration	
Licenses and registrations required for: Contractor to operate in country: Local incorporation: Capitalisation: Joint ventures rules: Local partner rules: Other:	Licences and registrations required for planning and construction operations: Planning: Construction: Environmental Impact: Health and safety: Other:

5. GENERAL

5.1 Special Considerations	
Any special contractual provisions required because of potential local problems:	Special incentives necessary to attract good quality expatriate managers and tradesmen:
Extent of national or local holidays: Dates: Mandatory/Optional** **delete as necessary	Local pressure group:

5.2 Physical Conditions at Site **delete as applicable		
Nature of soil Nature of subsoil: Bearing Capacity – Contractor commissioned survey report attached: Yes/No**	Natural Drainage:	Site Clearance Issues:
Location of Nearest Survey Monument:	Disposal of Excess Soil – Nearest Tip: Cost of disposal:	Nearest Location of Fill Material: Fill Material Cost: Availability:

5.3 Photographs Taken **delete as applicable		
Proposed Site: Yes/No**	Immediate Vicinity of Site: Yes/No**	Site Storage Areas: Yes/No**
Airport: Yes/No**	Railway Station: Yes/No**	Port: Yes/No**
Port Storage Facilities: Yes/No**	Camp Area: Yes/No**	Nearest Town: Yes/No**

5.4 Climate			
Month:	**Average Rainfall (mm):**	**Average Temperature (C):**	**Average Humidity(%):**
January		High:............... Low:.............	
February		High:............... Low:...........	
March		High:............... Low:...........	
April		High:............... Low:...........	
May		High:............... Low:...........	
June		High:............... Low:...........	
July		High:............... Low:...........	
August		High:............... Low:...........	
September		High:............... Low:...........	
October		High:............... Low:...........	
November		High:............... Low:...........	
December		High:............... Low:...........	
Maximum Storm Rainfall in 24 hour period: ....mm during.....................(month)		Maximum Storm Rainfall in 1 hour period: ....mm during..................(month)	
Average Wind Velocity............... .km/hr .... days/year		Average Storm Wind............... .km/hr .... days/year	
Typhoon/Hurricane/Cyclone Seasons: (between which months):		Rainy Season(s): (between which months):	

5.5 Temporary Access Roads, Storage Facilities	
Temporary works needed for construction access to be provided by Employer:	Temporary works needed for construction access to be provided to be provided by Contractor:
Temporary works to be incorporated into permanent works upon completion of Contract:	Location and size of temporary storage yard provided by Employers:
Availability of other storage facilities in area if necessary: Location: Size: Cost:	 Location: Size: Cost:

5.6 Accessibility of Site to Main Transport Links		
Nearest paved road: Category: Condition:	Nearest railroad track: Condition: Level of current usage:	Nearest railway station:
Nearest suitable sea port:	Nearest suitable river port:	Nearest airport:

5.7 Services ** delete as applicable		
Nearest supply source of potable water:	Nearest supply source of construction water: Pipe: Well:	Rates for water: Payable to:
Water treatment required: Yes/No**	Source of ice supply: Price: Quality:	Construction power to be supplied by: Contractor/Employer** Mains/Generator/Both**
Mains Supply Power : Reliability: Price:	Notice required to tap into existing power source:	Will drop down transformer be required: Yes/No**
Responsibility for provision of transmission lines to site boundary: Employer/Contractor**	Other power related issues:	Mains Gas: Available: Yes/No** Reliability: Price:
Other gas related issues:	Location of main sewer tie-in:	If no main sewer: Tanker availability: Cost of Tanker: Cost of disposal:
Location of nearest major hospital:	Location of nearest emergency/ casualty treatment centre:	Availability of medical facilities for expatriate workers:

Doctor to be located on Site: Yes/No** Full Time/Part Time	Nurse to be located on Site: Yes/No** Full Time/Part Time: Number of Nurses	Ambulance to be located on Site: Yes/No** Full Time/Part Time

5.8 Road Communication **delete as applicable		
Distance to port (km):	Distance to nearest major town (km):	Load capacity of road bridges on route: Adequate/Need Strengthening**
Traveling restrictions:	Permits required in respect of: excess load: Yes/No** excess height: Yes/No** excess weight: Yes/No**	Issues regarding damage to highways: Contractor to repair** Contractor to pay damages** Other**
Road haulage charges:	Special equipment availability:	Road tolls:

5.9 Railroad		
Capacity: Load restrictions:	Frequency of services:	Rates for rail transportation: Rates for demurrage:
Rates for switching to spur tracks:	Unloading facilities: Rates for unloading:	Rail permits required:

5.10 Unloading Quay **delete as applicable		
Unloading facilities: Weight limits:	Dredging: Necessary/unnecessary** Responsibility of Contractor/ Employer/Port**	Check any special requirements:

5.11 Telephone/Email Service		
Availability of lines: Waiting time: Limit on number of lines:	Rates for: Business lines: Private lines:	Radio telephone feasibility: Yes/No** Necessary/not necessary**:
License required for radio phone: Yes/No**	Internet access available: Yes/No**	ISP's rates:

5.12 Mail **delete as applicable		
Air-mail service availability: Yes/No**	Mail transit time to head office: Mail transit time from head office	Limitations of service:
Courier service availability: Yes/No**	DHL; TNT; Fed-Ex; UPS (circle if available)	Other:

6. MATERIALS AVAILABLE LOCALLY[8]

6.1 Civil Materials **delete as applicable		
Crushed Rock for concrete aggregate Screening: Cleanliness: Type of stone: Availability per day: Distance from site: Prices:	**Sand for concrete:** Salt content: Cleanliness: Quality: Availability per day: Distance from site: Price per m3:	**Cement:** Quality: Availability per day: Price per bag:
Timber: Locally available woods: Compare with US/EU known varieties Quality of seasoning: Accuracy of cutting: Availability: Costs:	**Rebar:** Availability: Price per kg: Price for bending to Specifications and drawings:	**Ready Mixed Concrete:** Maximum availability per day: Price per m^3: Supplier's own transit mixers Yes/No** Contractor to supply mixers Yes/No** Location of nearest test laboratory: Costs of concrete testing per cube:
Structural Steel: Quality of local products: Availability to specified standards: Prices of various sections: Import restrictions/tariffs: Price per tonne for fabrication service:	**Steel and Cast Iron Pipes**: Quality of local products: Availability to specified standards: Prices of various specifications and sizes:	**Reinforced Concrete, Concrete and Vitrified Clay Pipes:** Quality of local products: Availability to specified standards: Prices of various sizes:

6.2 M & E		
Quality of local cable products: Availability to specified standards: Prices of various specifications: Import and tariff restrictions:	Quality of local equipment & service: Availability: Maintenance locally: Spare parts availability: ASME, API etc. compliance: Guarantee terms, prices etc:	Prices for raw pipes: Prices for spooling pipe: Prices for cables: Availability of supply: Availability of: x-ray services: stress relieving services: sand blasting services: painting facilities:

7. LOCAL SUB-CONTRACTORS[9]

List of available sub-contractors in various disciplines with details of financial standing and local reputations for standard of workmanship, reliability, productivity, schedule control etc. (List to be attached)	Guideline prices for principal sub-contract works (ground works, substructures, piling, dredging, concrete frames, pipe supports, light/heavy steel structures, buildings, external works, etc. (List to be attached)	Standard forms of contract and sub-contract in common usage in the site vicinity. (List and samples of contract forms to be attached)

8. SHIPPING AND IMPORTATION

8.1 Port Facilities		
Nearest suitable port:	Tonnage handled per year:	Limitations of weight: Limitations of bulk:
Storage space available:	Nearest alternative port:	Demurrage charges:

8.2 Port Expenses		
Cargo handling:	Storage:	Local Agents:
Port clearance:	Customs clearance:	Other special costs:

8.3 Import Licences **delete as applicable		
Extent of Contractor's responsibility:	Extent of Employer's responsibility:	Use local agent: Yes/No** Employ local staff Yes/No**

8.4 Customs Requirements		
Extent of inspections:	Time taken for clearance:	Paperwork required by Customs:

8.5 Customs Duties **delete as applicable		
Tariff classification and rates tables: Attached: Yes/No**	Special duty concessions: Preference/most favored nation/ other** (if 'other' attach detailed information)	Project specific concessions: Degree to which they apply:

8.6 Air Freight **delete as applicable		
Nearest Domestic airport: Name: Location: Size:	Nearest International airport: Name: Location: Size:	Flight schedules attached: Yes/No** Typical freight rates attached: Yes/No**
Freight through-put per month:	Special arrangements for quick clearance of rush deliveries: Details attached: Yes/No**	Intermediate freight transshipment point between point of embarkation and delivery airport: Yes/No** Details:

8.7 Rail & Ocean Freight **delete as applicable		
Rates for crated rail freight: Attached: Yes/No** Rates for uncrated rail freight: Attached: Yes/No**	Rates for sea freight above deck: Attached: Yes/No** Rates for sea freight below deck: Attached: Yes/No**	Heavy lift rates: For rail freight – attached: Yes/No** For sea freight – attached: Yes/No**

9. CONSTRUCTION EQUIPMENT AND SUPPLIES ** delete as applicable

Local availability: List of rental companies attached: Yes/No** Rates of local rental: Rates lists attached: Yes/No**	Imported equipment can be: Re-exported: Yes/No** Sold locally: Yes/No**	Sale of imported equipment: Duties: Taxes: Other:
Local purchase prices: Price lists attached: Yes/No**	Availability of spare parts: Excellent/Good/Poor**	Availability of repair facilities: Excellent/Good/Poor**
Availability of consumables: Excellent/Good/Poor**	Local agents of known makes: List attached: Yes/No**	

10. LOCAL LABOR COSTS AND PRODUCTIVITY **delete as necessary

Direct Hire: Yes/No** Agency: Yes/No** Both: Yes/No**	Local wages per craft: List attached: Yes/No**	Local non-craft wages: List attached: Yes/No**
Public Holidays to be paid: Yes/No** Religious holidays to be paid: Yes/No** Statutory holidays to be paid: Yes/No** Other:	Employee transportation requirements: Company transport: Own transport: Both:	Medical care requirements: Insurance: Hospital registration: Other:
Statutory benefits to be paid:	Additional benefits usual in area:	Trade union situation:
Productivity norms: Summary attached: Yes/No**	Comparison of local productivity norms with similar norms in US/EU: Summary attached: Yes/No**	On-site catering requirement:
Availability, salaries etc. for local administrative staff: List attached: Yes/No**	Salaries, benefits and incentives etc. for non-manual expatriate employees: List attached: Yes/No**	

11. FOREIGN WORKERS AND SPECIALISTS **delete as necessary

Site organisation: Tables & charts attached/To be prepared later**	Local laws governing ratio of foreign employees to locals: Attached: Yes/No**	Immigration visa requirements, limits of use and limits of stay, application time: Details attached: Yes/No**
Labor permit requirements: Details attached: Yes/No**	Health clearance requirements for expatriate employees: Details attached: Yes/No**	Insurance requirements for expatriate employees: Details attached: Yes/No**
Inoculation requirements: Typhoid/Hepatitis/Cholera/ Rabies/Yellow Fever/Other**	Local costs of living (food, services etc.): Details attached: Yes/No**	Local education facilities for expatriate children – where located, costs: Details attached: Yes/No**

12. CAMP FACILITIES **delete as applicable

Available local accommodation, with costs, rental conditions (deposits etc.): Details attached: Yes/No**	Temporary accommodation be turned over to Employer on completion: Yes/No**	Suitable furniture and furnishings available locally: Yes/No**
Food facilities in camp: Canteen for workers: Yes/No** Restaurant for management: Yes/No** Supermarket/shop: Yes/No**	Recreational facilities in camp: TV Lounge: Yes/No** Swimming Pool: Yes/No** Squash Courts: Yes/No** Tennis Courts: Yes/No** Gymnasium: Yes/No**	Living accommodation in camp: Dormitory: Yes/No** Single Units: Yes/No** Double Units: Yes/No** Family Units: Yes/No**
		Availability of local catering service:
Laundry service to be provided: Yes/No**	Room cleaning service to be provided: Yes/No**	Local laundries: Yes/No** Local cleaners: Yes/No** Details attached: Yes/No**

1. The nature and extent of this typical form can be changed to suit each particular project, depending upon the size and complexity of each project.
2. Obtain location map of Site and nearest city/town and detailed survey map of Site location.
3. Long term risk issue, could affect financing package.
4. Could affect financing package.
5. Could affect site security, labour and sub-contractor availability and attitude.
6. *ibid.*
7. Local media can have a significant influence upon the success or failure of a project, particularly in less well developed countries.
8. Obtain business cards from local company representatives and copies of local telephone directories.
9. *ibid.*

Appendix 4

Contractor's Pre-Tender Information Confirmation[1]

Either:

** We confirm that we have not carried out a Pre-Tender Site Visit and we further confirm that all factors affecting our Tender, including but not limited to the following items numbers 1 to 12 inclusive, have been duly considered and taken into account in the preparation of our Tender.

Or:

** We confirm that we visited the Site on …… day of…………….20 … and met with the representatives of the Employer and all other persons whom we, in our sole judgment, deemed it necessary to meet with at that time. We confirm that the following Items 1 to 12 were discussed and agreed and that any and all matters and considerations in relation thereto arising from our meetings and discussions, and which we have or should otherwise have become aware of affecting or likely to affect our Tender, have been duly considered and taken account of in the preparation of our Tender whether or not specifically listed below.

**** Delete as appropriate**

1. Access to and within the Site together with rights of way requirements and any special restrictions on movement of materials, equipment or vehicular traffic as regards route, weight or size and restrictions on access and movement of personnel etc., including temporary roads, access ways, barriers etc. to be provided by Contractor and to be provided by others.
2. Availability and quality of local labour, local labor agreements, rates of pay, allowances, normal site working hours, overtime restrictions and requirements, union relationships, site industrial relations procedures.
3. Availability, cost and quality of local plant, equipment, material.
4. Conditions, location and nature of the existing works and facilities (where applicable) including roads, bridges, pipe bridges, pipelines etc. within and adjacent to site, that need to be taken account of during the execution of the Contract.
5. Employer's, works, statutory and local regulations including Health, Safety and Environmental regulations, and operational practices and procedures manuals currently in force.
6. Details of other contractors working on the Site during the execution of these Contract works.
7. Material storage facilities to be provided by Contractor and to be provided by others, including type, size, location, availability, and material receiving/distribution and control procedures to be followed.
8. On/off site car, cycle and bus parking facilities, permits, entry requirements and restrictions, site speed limits etc. for site staff, workers and visitors.
9. Position of supplies and availability of temporary electrical power, lighting and water services.
10. Procedures for disposal of spoil, rubbish and waste as appropriate and relevant, and restrictions thereto, including contaminated and toxic or otherwise hazardous waste.
11. Site accommodation to be provided by Contractor and provided by others, including size, location, responsibility for provision of services etc. and responsibility for the provision of Site telephones, radios, walkie-talkies and other means of site communication to be used by Contractor's personnel.

12. Site location and conditions, including soil, geological, environmental and climatic conditions, and location of site bench marks.

Signed for and on behalf of the Contractor	Signed for and on behalf of the Employer
....	....

1. Signature of this form will limit the possibility of a claim for lack of or incorrect pre-contract information should problems be encountered during the execution of the work, giving comfort to the employer and encouraging the contractor to carry out adequate pre-contract investigations. Similar forms can be used by the contractor in its negotiations with prospective sub-contractors.

Daywork Record Sheet

[CONTRACTOR'S COMPANY NAME]	DAILY DAYWORK RECORD SHEET

Date:	Site Instruction/Purchase Order Number:
Project Title:	Contract No.:
Employer:	Daywork Record Reference Number:

DESCRIPTION OF WORKS TO BE EXECUTED ON DAYWORK

LABOUR

Name	Trade:	Standard Hours:	Overtime Hours:

PLANT/EQUIPMENT:

Description:	Standard Hours:	Overtime Hours:

MATERIALS:

Description:	Quantity:

AUTHORISATION:

Issued for Contractor: Signed:	Position:	Date:
Accepted by Employer: Signed:	Position:	Date:

Appendix 6:

Overage, Shortage and Damage (OSD) Report[1]

Project Title:	Contractor:
Client:	Contract No:
OSD Report No:	Report Originated by:
Shipment No:	Inspection Location:
Description of Items Shipped:	Vessel Name/Number:

Purchase Order No:		Package No:		Vendor Name:
No. of items on PO:	Billed Quantity:	Actual Quantity:	Description:	Remarks:

Carriage Method: Ship/Air/Truck/Hand	Carrier Company Name:	Carrier Co. Reference No:
Container Type:	Container Damage Details:	Container Stowage Location: On Deck/Below Deck
Inspection Release Certificate Supplied: Yes/No Date:	Delivery Dispatch Notice Endorsed: Yes/No Date:	Carrier Notified: Yes/No Date:
Insurance Co. Survey Report Attached: Yes/No Report No: Date:	Insurance Incident Report Attached: Yes/No Report No: Date:	Date Shipment Received: Date Shipment Inspected:

Repair/Replacement Available Locally: Yes/No If Yes, details:	Estimated Local Costs of:		Remarks:
	Repair/ replacement:	Delivery:	

Signed:.. (on behalf of Main Contractor)	Signed:.. (on behalf of Insurance Company)
Name in Capitals:......................................	Name in Capitals:......................................
Company:...	Company:...
Position:............................ Date:..............	Position:............................ Date:..............

1. An OSD Report is discussed in the 'Fundamental Elements of Effective Contract Management' section starting on p.127.

Appendix 7

Progress Report No ..... for the Month of ....

Project Title

Table of Contents[1] **Page No.**

Section I **Executive Summary**

General
Design Engineering
Procurement
ite Construction (Civil Progress)
Commissioning and Tests on Completion
Project Overall Progress
Site Safety Record Summary
Document Record Summary
Site Manpower and Equipment Record Summary
Outstanding Employer Actions and Decisions
Milestone Summary Schedule

Section II **Project Management and Project Status**

A. Project Milestone Summary
B. Contract Programme Status
1.1 Total Project Percent Complete
1.2 Modified Contract Programme Update
1.2.1 Narrative
1.2.2 Modified/Current Programme
1.2.3 Current Month: Actual v Planned
1.2.4 Current Programme v Contract Programme
1.3 Three month look-ahead programme

Section III **Scope of Work and Current Status**

A. Design Engineering
1.1 Drawing & Specification Status
1.1.1 Drawing Status
1.1.2 Specification Status
1.2 Overview
1.3 Architectural
1.4 Civil/Structural
1.5 Control Systems
1.5 Electrical
1.6 Mechanical
1.7 Plant Design
1.8 Switchyard
1.9 Engineering Documents Outstanding/Overdue
B. Material/Equipment Procurement
1.1 Material

1.1.1 Orders placed during month
1.1.2 Orders scheduled to be placed next month
1.2 Equipment
1.2.1 Orders placed during month
1.2.2 Orders scheduled to be placed next month
C. Manufacturing/Shipping
1.1 Manufacturing
1.2 Shipping
1.3 Customs Clearance, Duties etc.
D. Testing/Inspection
1.1 Testing
1.2 Inspection
1.2.1 Inspection Status
1.2.2 Non-conformance Reports
E. Construction/Erection
1.1 Construction
1.2 Erection
F. Commissioning/Tests on Completion
1.1 Commissioning
1.1.1 Progress to Date
1.1.2 Progress Planned for Next Month
1.1.3 Progress Planned to Completion
1.2 Tests on Completion
1.2.1 Progress to Date
1.2.2 Progress Planned for Next Month
1.2.3 Progress Planned to Completion
G. Material/Equipment Inspection Report
1.1 Inside Country
1.2 Outside Country
H. Start-Up/Operator Training
1.1 Start-Up
1.1.1 Progress To Date
1.1.2 Progress Planned for Next Month
1.1.3 Progress Planned to Completion
1.2 Operator Training
1.2.1 On-Site
1.2.1.1 Classroom Training
1.2.1.2 Practical Training
1.2.2 In Factory
1.2.2.1 Classroom Training
1.2.2.2 Practical Training
I. Vendor Pre-Qualification Status

Section IV **Contract Changes/Clarifications**

A. Change Order Proposals Log
B. Technical Change Notices Log
C. Proposed Specification Changes Log
D. Requests for Information (RFI) Log

Section V **Commercial and Administrative Matters**

A. Payment Status
  1.1 Payment Requests Accepted
  1.2 Payment Requests Paid
  1.3 Payment Requests Outstanding
  1.4 Payment Requests in Dispute
B. Earned Value Curves
C. Bonds – Status Report
D. Statutory Approvals/Licenses Report
E. Insurance
  1.1 Insurance Policy Status Report
  1.2 Insurance Claims Status
    1.2.1 Insurance Claims Submitted
    1.2.2 Insurance Claims Settled
    1.2.3 Insurance Claims Rejected
    1.2.4 Insurance Claims Under Discussion
    1.2.5 Insurance Claims Outstanding
F. Risk Log
G. Back-Charge Status
  1.1 Back-Charges Submitted
  1.2 Back-Charges Settled
  1.3 Back-Charges Rejected
  1.4 Back-Charges Under Discussion
H. Claims Status
  1.1 Claims Submitted
  1.2 Claims Agreed
    1.2.1 Agreement In Principle
    1.2.2 Commercial Agreement
  1.3 Claims Rejected
  1.4 Claims Under Discussion
  1.5 Claims Outstanding
I. Documentation
  1.1 Correspondence During Month
    1.1.1 Employer/Employer's Representative to Contractor
    1.1.2 Contractor to Employer/Employer's Representative
    1.1.3 Other Significant Correspondence
  1.2 Minutes of Meetings

1.2.1 Regular Progress
1.2.2 Technical
1.2.3 Commercial
1.2.4 Schedule
1.2.5 QA/QC
1.2.6 Commissioning
1.2.7 Start-Up
1.2.8 Punch List
1.2.9 Warranty
1.2.10 Other

Section VI **Site Safety**
A. Incident Report for Month
B. Lost Time Record
C. Medically Treated Injuries Record
D. Incident-Free Days Log
E. Safety Record to Date
F. Safety Training Undertaken During Month

Section VI **Progress Photographs**

1. The Table of Contents used here is a typical listing for a major engineering project, such as a power station. It is intended to give the reader an example of a workable format that can be adapted to suit a particular project. The format/list of contents should be specified in the contract documentation so that the employer can ensure that he will receive all the relevant information that it considers necessary for project planning, project financing and reporting on progress to boards, banks, financial institutions and their professional advisers.

   The contractor will find its compilation a useful exercise and will benefit from the discipline that the preparation and maintenance of necessary records will impose on its staff.

   Both parties will benefit should a dispute situation arise at a later stage of the contract since they will have a ready source of records upon which to draw.

Change Order Proposal[1]

Project Title:	Employer's Name:
Contractor's Name:	Proposal Reference No.:
Subject of Proposal:	Date of Proposal:
Detailed Description of Proposal:	
Relevant Documents:	Estimate of Value of Change Order Proposal:
Effect on Completion Date: None/To be Assessed/As follows:** …………………………………………………….. **Delete as necessary	Effect on Contract Price: None/To be Assessed/As Follows:** ………………………………………………… **Delete as necessary
Other Effects:	
Issued on behalf of Contractor by: Signature:…………………………………. Name:……………………………………. Position:………………………………….. Company:……………………………….. Date:……………………………………….	Accepted on behalf of Employer by: Signature:…………………………………. Name:……………………………………. Position:………………………………….. Company:…………………………………. Date:……………………………………….

1. This can also be referred to as 'Variation Order Proposal' depending upon the terminology used in the contract. A similar form could be used by the sub-contractor in relation to the contractor.

Appendix 9

Sub-Contractor Taking Over Certificate[1]

Contractor's Name:	**Client/Employer's Name:**
Project Title:	**Project No:**

Client/Project Owner:	Certificate No:
Sub-Contractor:	Date of Issue of Certificate:
Sub-Contract Works:	Completion/Taking Over Date:
Whole of Sub-Contract Works Taken Over: Yes/No**	Start Date of Defects Liability Maintenance Period: ....
	End Date of Defects Liability Maintenance Period: ....
Section of Sub-Contract Works Taken Over: Yes/No** If Yes, Section:...	Start Date of Defects Liability Maintenance Period:....................................
	End Date of Defects Liability Maintenance Period: ....

We hereby certify that the Section/Whole** of the above Sub-Contract Works have been completed in accordance with the Sub-Contract (except for Minor Defects and Minor Outstanding Works detailed below which do not affect the use, operation or occupancy of the Facility/Works**) and have passed the Tests on Completion and that the Section/Whole** of the Sub-Contract Works shall be deemed Taken Over on the date shown above and the Defects Liability Period/Maintenance Period shall commence from and end on the dates shown above.

This Taking Over Certificate does not relieve the Sub-Contractor of his obligations for remedying the Minor Defects and completing the Minor Outstanding Works listed below, for remedying all other defects and deficiencies and for performing and completing the Works in accordance with the Sub-Contract.

Minor Defects and Minor Outstanding Works	
Section/Area:	Description:

Issued on behalf of Main Contractor by:	Accepted on behalf of Sub-Contractor by:
Signature: .... Date............	Signature: .... Date............
Name:..	Name:..
Position:..	Position:..
Company:..	Company:..

**** Delete as necessary**

1. Possibly better known in the UK as 'Practical Completion Certificate'; also referred to as 'Substantial Completion Certificate'.

Sub-Contract[1] Close Out Check-List

Contract Title:	**Client:**	**Contractor:**
Sub-Contractor: Company Name: Contact Person:	Sub-Contract Works: Description:	Sub-Contract Order No: {Fixed Lump Sum} {Remeasurable} {Daywork}** **Delete as necessary
Order Value:	Final Account Value:	
Commencement Date:	Completion Date:	Delay (Days):
Extensions of Time Granted: Days:	Prolongation Costs Agreed:	Liquidated Damages Deducted:

Check List Items:	Yes	No	N/A	Signature
1. Final Inspection Carried Out				
2. Documentation Received:				
a) as-built drawings				
b) operation manuals				
c) maintenance manuals				
d) training certifications				
e) material certificates				
f) certificates of origin of material/equipment				
g) QA/QC Dossier (weld records, test certificates, etc.)				
h) performance test results				
i) guarantees and warranties				
3. Free issue materials:				
a) reconciliation statement agreed				
b) surplus materials handed back				
4. Acceptance/Completion Certificate issued				
5. All sub-contract amendments/change orders carried out				
6. Insurance claims settled				
7. OS & D claims settled				
8. Liquidated Damages settled				
9. Back charges paid by/deducted from payments due to sub-contractor – resolved				
10. Final Account agreed to and signed off by sub-contractor				
11. All disputes, claims, mediations, arbitrations etc. resolved				

Appendix 10: Sub-Contract Close Out Check-List

Sub-Contract Close-Out Check[2] List continued:

*(**Note:** The preceding sheet will be initiated by the first person in the Contractor's organisation on the following list and will then be routed through the other persons listed in the order of listing, each person completing that portion relevant to his department and initialing accordingly, before returning to the initiator, fully completed.)*

Circulation Route:	**Signature:**
Contract Manager:	....
Site Manager:	....
Procurement Manager:	....
Design Manager:	....
Materials Controller:	....
QA/QC Manager:	....
Commissioning Manager:	....
Project Controls Manager:	....
Quantity Surveyor/Cost Engineer:	....
Contract Manager:	....

Approval of Sub-Contract Close-out:

This is to certify that Final Certificate in respect of the Sub-contract to which this Close-out Check List refers was issued and the Final Invoice was approved for payment on .... day of .... 200....

Approved By:	**Verified and Agreed By:**
Signed:..	Signed:......................................
Name:..	Name:......................................
Position: Contractor's Project Manager	Position: Sub-Contractor's Project Manager
Date:..	Date:......................................

Distribution:

(as required by particular Company regulations)[3]

1. A similar check-list format can be used between the employer and the contractor where the nature of the project allows.
2. The term 'Close Out Check-list' might not be familiar to many who have not had international experience but is an easily recognisable principle. 'Close Out' is basically the process of ensuring that all the sub-contract requirements have been met before finalising the sub-contractor's account. This is an essential step which must be taken on international contracts in particular where there is more scope for confusion because of language difficulties, contract interpretation issues, confusion in amalgamating local and specific contract practices and where the stakes are often higher than in a domestic contract. See separate section 'Contract Close Out' on p.66.
3. This form could be used, subject to the employer's agreement, where the main contract calls for the contractor to satisfy the employer that all sub-contractor accounts have been settled and no liens exist as a condition precedent to issuing the contractor's final certificate or payment.

Sub-Contractor Final Completion Certificate[1]

Contractor's Name:	Employer's Name:
Project Title:	**Project No:**

Certificate No:	Date of Issue of Certificate:
Sub-Contractor:	Taking Over Date:
Sub-Contract Works:	
Whole of Sub-Contract Works Finally Completed: Yes/No**	Start Date of Defects Liability Maintenance Period: .... End Date of Defects Liability Maintenance Period: ....
Section of Sub-Contract Works Finally Completed: Yes/No** If Yes, Section...	Start Date of Defects Liability Maintenance Period: .... End Date of Defects Liability Maintenance Period: ....
We hereby certify that the Section/Whole** of the above Sub-Contract Works have been completed in accordance with the Sub-Contract and that the Defects Liability Maintenance Period for the Section/Whole** of the Sub-Contract Works shall be deemed to have ended on the date shown above.	
Issued on behalf of Main Contractor by: Signature: .... Name: .... Position: .... Company .... Date:...	Accepted on behalf of Sub-Contractor by: Signature: .... Name:... Position:...................................... Company .... Date:..

**** Delete as necessary**

1. This can also be referred to as the 'Sub-Contractor's Making Good Defects Certificate' depending upon the terminology used in the sub-contract.

Law Report Abbreviations & Table of Cases

Law Report Abbreviations (as used in this book)

AC	Law Reports Appeal Cases	**F.**	Session Cases, 5th. Series [Fraser] (Scotland) 1898 to 1906
ALJR	Australia Law Journal Reports	**HL**	House of Lords
All ER	All England Law Reports	**Ill.**	Illinois Reports (US)
App. Cas	Law Reports, Appeal Cases	**JP**	Justice of the Peace / Justice of the Peace Reports
BCC	British Company Law Cases	**KB**	Law Reports: Kings Bench Division
BCL	Building and Construction Law (Australia)	**LGR**	Local Government Reports
BCLC	Butterworth's Company Law Cases	**LR Exch.**	Law Reports: Exchequer 1st series 1865–75 (Volume No. inserted after LR)
Bing.NC	Bingham's New Cases (BNC)	**LT**	Law Times Reports
BLR	Building Law Reports	**Lloyd's Rep.**	Lloyd's List Law Reports
CA	Court of Appeal	**Mass.**	Massachusetts Reports (US)
Ch	Law Reports: Chancery Division 1991	**NE**	Northeastern Reporter (US)
Ch. D	Law Reports – Chancery Division 1875–90	**NSR**	Nova Scotia Law Reports
CILL	Construction Industry Law Letter	**NTSC**	Northern Territories Supreme Court (Australia)
CL	Current Law	**NZLR**	New Zealand Law Reports
CLJ	Cambridge Law Journal	**PD**	Law Reports: Probate
Co. Rep	Coke (K.B. Series 1572 to 1616)	**QB (or QBD)**	Law Reports: Queen's Bench Division
ConLR	Construction Laws Reports	**SA**	South African Law Reports
Const. LJ	Construction Law Journal	**Salk.**	Salkeld (K.B. Series 1689 to 1712)
CP	Law Reports: Common Pleas 1875-80	**Scots CS**	Scottish Court of Sessions
DLR	Dominion Law Reports	**SLR**	Singapore Law Reports
EG	Estates Gazette	**TCC**	Technology and Construction Court
EWHC	High Court (England and Wales)	**TLR**	Times Law Reports
Ex. (or Exch.)	Exchequer	**WLR**	Weekly Law Reports
F.	Federal Reporter (US)		

Table of Cases

The page number is the page from which reference to the footnote is made. The footnote number is given in superscript after the page number.

Case	*Page*
A. B. Contractors Ltd v Flaherty Brothers Ltd (1978) CA	226[4]
Actionstrength v International Glass Engineering (3 April 2003), HL	143[6]
Ailsa Craig Fishing Co. Ltd v Malvern Fishing Co. (1983) 1 WLR 964	115-16[7]
Amalgamated Building Contractors Ltd v Waltham Holy Cross UDC (1952) 2 All ER 452C	194[1]; 239[4]
Aoki Corp. v Lippoland (Singapore) Pte. Ltd (1995) 2 SLR 609	8[3]
Ascon Contracting Ltd v Alfred McAlpine Construction Isle of Man Ltd (1999) 66 ConLR 119	123[3]
Aspeden v Seddon (No. 2) (1876)	24[2]
Attorney General for the Falkland Islands v Gordon Forbes Construction (Falklands) Ltd (No. 2) (2003) (LTL 14/3/2003)	62[3]
Aurum Investments Ltd v Avonforce Ltd (in liquidation) (2001) CILL 1729	106[8]
BP Refinery (Westernport) Property Limited v Shire of Hastings (1978) 52 ALJR 20	137[1]
B. Hargreaves Ltd v Action 2000 Ltd (1992), CA	225[1]
B. Mullen and Sons (Contractors) Ltd. v Ross and Another (1995)	145[3]
Banque Quilpue Ltd v Bryant (1904)	194[3]; 238[1]
Bim Kemi v Blackburn Chemicals Ltd (3 April 2001), CA	225-6[2]
Birkmyr v Darnall (1704) 1 Salk. 27	142[2]
Blinderman 69 F. (2d) 552 (1982), US Court of Appeals for the Federal Circuit	55[5]
Blyth & Blyth v Carillion Construction Co. Ltd (18 April 2001)	26[9]
Bolam v Friern Hospital Management Committee (1957)	201[1]
British & Commonwealth Group Plc. v Quadrax Holdings Inc. (1983) 3 All ER 492	244-5[5]
British Westinghouse Electric & Manufacturing Co. Ltd v Underground Electric Railways of London Ltd (1912) AC 673	175[3]

Brunswick Construction Ltd v Nowlan (1974) (Canada) 105[3]

Bunge Corporation v GATX Corporation – US – Louisiana Supreme Court 108[11]

CCC Films (London) v Impact Quadrant Films (1985) QB 16 76[8]

Cargill v Bangladesh Sugar and Food (1998) 2 All ER 406 29[2]

Central London Property Trust Ltd v High Trees House Ltd (1947) KB 130 58[11]

Chaudry v Prabhakar (1989) 201[2]

Chillingworth v Esche (1924) 1 Ch. 97 166[3]

City Inn v Shepherd Construction Ltd (2003) Scots CS 146 240[8]

Clayton Love v B & I 116[9]

Compagne Financiere du Pacifique v Peruvian Guano Company (1882) 11 QBD 99[12]; 100[14]

Confetti Records and Others v Warner Music UK Ltd (2003) EWHC 1274 228-9[2]

Costain Civil Engineering Ltd and Tarmac Construction Ltd v Zanen Dredging and Contracting Co. Ltd (1996) 85 BLR 85 197[2]

Courtney & Fairburn Limited v Tolaini Brothers (Hotels) Limited (1975) 1 All ER 716 165[1]

Currie v Misa (1875) LR 10 Ex. 153 57[8]

DMA Financial Solutions Ltd v Baan UK Limited and Others (28 March 2000), ChD 228[1]

D. O. Ferguson Associates v Sohl (1992) 62 BLR 95 198 206[4]

Day v Mead [1987] 2 NZLR 443, CA 54[5]

Doiron v Caisse Populaire D'Inkerman Ltee (1985) 17 DL. (4th) 660 54[4]

Don King Productions Inc. v Warrant (1999) 2 All ER 218 25[5]

Dudley Corporation v Parsons and Morrin Ltd (1959) 235[2]

ENS Limited v Derwent (1998) 62 ConLR 141 29[1]

Edward Lindenberg v Joe Canning Jerome Contracting Ltd (1992) 105[2]; 106

Embankment Place Hotels (Blackfriars) Limited v (1) Blackfriars Hotels Limited; (2) Six Continents Hotels (UK) Limited (formerly Holiday Inns (UK) Limited) (2002) 187[4]

Equitable Debenture Assets Corporation Ltd v William Moss & Others (1984) 2 ConLR 1 105[6]

Europools Plc v Clydeside Steel Fabrications Limited (17 January 2003) (Outer House Court of Session, Scotland) 78[11]

Felton v Wharrie (1905) 245[8]

Fibrosa Spolka Akcjna v Fairbairn Lawson Combe Barbour Ltd [1943] AC 32; Lord Wright 205[2]

Fisher v Ford (1840) 239[5]

Freeman & Son v Hensler CA (1900) 64 JP 260 10[1]

Freeman v Marshall & Co. (1966) 200 EG 777 201[2]

G H Myers & Co. v Brent Cross Service (1934) 1 KB 46 119[1]

Galbraith v Grimshaw and Baxter (1910) 1 KB 339 147[8]

Gaymark Investments Pty Ltd v Walter Construction Group Ltd (1999) 16 CL 449; NTSC 143 239-40[6]

Grant v Cigman (1996) 2 BCLC 245[9]

Greaves (Contractors) Ltd v Baynham Meikle & Partners (1975) 3 All ER 99 119-20[3]

H v Schering (1983) 1 All ER 849 62[4]

Hadley v Baxendale (1854) 9 Ex. 341 75[1, 2]; 77

Henry Boot Construction Ltd v Alstom Combined Cycles Ltd (2000) BLR 247, CA 54[2]

Henry Boot Construction (UK) Ltd v Malmaison Hotel (Manchester) Ltd [1999] 70 ConLR 32, QBD (TCC) 235[1]

Henry Kendall & Sons v William Lillico & Sons (1969) 114[3]

Heydon's case (1584) 3 Co. Rep. 7a 72[11]

Investor's Compensation Scheme v West Bromwich Building Society (1998) 1 All ER 98 69[4]

James Miller v Whitworth Street Estates (1970) AC 583 22[5]

Jarvis v Harris (1996), CA 142[4]

John Barker Construction Ltd v London Portland Hotel Ltd (1986) 194[5]

John L. Haley Ltd v Dumfries and Galloway Regional Council (1988) 187[3]

John Lelliott (Contracts) Ltd v Byrne Bros (Formwork) Ltd [1992] 31 ConLR 89 119[2]

John Mowlem & Co. Ltd v British Insulated Callenders Pension Trust Limited (unreported) 120[5]

Kemper v Don Coleman, Jr., Builder, Inc No. 35, 172 (1987), CA 108[12]

Kier Construction v Royal Insurance (UK) Ltd (1992) 186[2]

Kitson Insulation Contractors Ltd v Balfour Beatty Buildings Ltd (1990) 197[3]

Lachlani v Destination Canada (UK) Ltd (1996) 198[6]

Laserbore v Morrison Biggs Wall Ltd (1992) 198[5]

Linden Gardens Trust Ltd v Lenesta Sludge Disposals Ltd (1994) 1 AC 85 25[4]

Lister v Forth Dry Dock and Engineering Co. Ltd (1989) 1 All ER 1134 72[12]

Liverpool CC v Irwin (1976) 2 All ER 39 138[3]

Lombard North Central Plc v Butterworth (1987) QB 527 244[1]

London Borough of Merton v Lowe (1981) 18 BLR 130 201[4]

Lucas v Godwin (1837) 3 Bing. NC 737 245[6]

Mobil Sales and Supply Corp v Pacific Bear (owners) [1979] Hong Kong 147[9]

Morrison–Knudsen Co. Inc. and Others v British Columbia Hydro and Power Authority (1991) 7 Const. LJ 227 8[2]

Mostcash Plc (formerly UK Paper Plc) v Fluor Limited (11 January 2002) (TCC) 69[4]

Motemtronic Ltd v Autocar Equipment Ltd (20 June 1996) (unreported), ChD 142-3[5]

Mouat v Clark Boyce [1992] 2 NZLR 559, CA 54[5]

OTM Limited v Hydronautics (1981) 165[2]

Obagi v Stanborough (Developments) Ltd TLR 15 December 1993 76[7]

Olley v Marlborough Court Ltd (1949) 1 KB 532 114[2]

Olympia & York Canary Wharf (No.2) (1993) BCC 159 244[4]

Peak Construction (Liverpool) Ltd. v McKinney Foundations Ltd (1 July 1970) 69 LGR 1; 1 BLR 111, CA 64[2]; 194[2]; 238[2]

Perini Corporation v Commonwealth of Australia (1969) 12 BLR 82 8[1]

Plant Construction Plc v Clive Adams Associates and JMH Construction ServicesLtd (20. December 1999) (unreported), CA; (31 March 2000), TCC 105[1]; 106

Pride Valley Foods Limited v Hall and Partners and Hall another Contract Management) Limited (28 June 2001), CA 105[7]

Ribic v Weinstein [1982] 140 DLR (3d) 258 54[4]

Roberts v Bury Commissioners (1870) LR 5 CP 310 10[2]

Robinson v Harman (1848) 1 Exch. 850 174[2]

Rotherham Metropolitan Borough Council v Frank Haslam Milan and Co. (1996) CLJ 333; 78 BLR 1 120-21[6]

Scarf v Jardine (1882) 7 App. Cas. 345 25[8]

Schwanbeck v Federal Mogul Corp. 412 Mass. 703 (1982) 168[10]

Simmons v Garces (2002) 198 Ill. (2d) 541; 763 NE (2d) 720 152[4]

Smith v Wilson (1832) 138[4]

Squire v Insurance Corp. of British Columbia (BCCA) (1990) 69 DLR 64[4]

State of West Virginia v Ottis Ray Euman No. 29700 70[6]

Stickney v Keeble (1915) AC 386, HL 245[7]

Strachan and Henshaw v Stein Industrie (8 December 1997), CA 112[1]

Strydom v Certain Underwriting Members (2000) (2) SA 482 (W) 64[3]; 116-17[11]

Surrey County Council v Bredero Homes TLR April 1993, CA 76[6]

Thomas National Transport (Melbourne) Pty. Ltd v May & Baker (Australia) Pty. Ltd (1966) 115 CLR. 353 116[8]

Teacher v Calder (1899) 1 F. (HL) 39 75-6[5]

The Moorcock (1886–90) All ER 530; 14 PD 64 138-9[5]

The Nanfri (1978) 1 QB 927 226[3]

The Royal Brompton Hospital NHS Trust v Frederick Alexander Hammond & Others (2002) (TCC) 54[3]

Tolhurst v Associated Portland Cement Manufacturers (1900) Ltd (1902) 2 KB 660 24[1]

Tompkins Hardware Ltd v North Western Flying Services Ltd (1982) 139 DLR (3d) 329 54[4]

Total Transport Corp. v Arcadia Petroleum Ltd (1998) 1 Lloyd's Rep. 357 141[1]

Trollope and Colls Ltd v Atomic Power Constructions Limited (1963) 1 WLR 333 — 167-8[7]

Trollope and Colls Ltd v North West Metropolitan Regional Hospital Board (1973) 2 All ER 260 — 194[4]

Turf Masters v TAG Developments and Dartmouth (1994) 135 NSR (2d) 105; Chapman J. A., Nova Scotia Court of Appeal — 198-9[8]

Turner Corporation Ltd. (Receiver and Manager appointed) v Austotel Pty. Ltd (1994) 13 BCL 378 — 240[7]

Twintec Limited v GSE Building and Civil Engineering Limited (2002) EWHC 605 (TCC) 24 March 2003 — 167[6]

United Scientific Holdings v Burnley Borough Council (1978) AC 904 — 244[3]

University of Glasgow v William Whitfield and John Laing (Construction) Ltd (1988) 42 BLR 66 — 107[10]

Victoria Laundry (Windsor) v Newman Industry Ltd (1949) 2 KB 528, CA — 75[3]

Victoria University of Manchester v Hugh Wilson & Lewis Wormersley and Pochin (Contractors) Ltd (1984) 2 ConLR 43 — 105[4]

Viking Grain v T. H. White Installations (1985) — 120[4]

Von Hatzfeld-Wildenburg v Alexander (1912) 1 Ch. 284 — 168[9]

Wells v Army & Navy Co-operative Society Ltd (1902) 86 LT 764 — 10[3]; 238[3]

Westdeutsche Landesbank Girozentrale v Islington LBC (1996) AC 669 — 205[3]

Whiteley v Chappell (1986) LR 4 QB 147 — 72[10]

Willmott v Barber [1880] 15 ChD 96 — 109[1]

Wimpey Construction (UK) Ltd v D. V. Poole (1984) — 201[3]

Wraight v P. H. & T. Holdings Ltd (1968) (unreported) — 189[1]

Yeoman Credit Ltd v Latter (1961) 2 All ER 294 — 142[3]; 142-3[5]

Index

Note: Footnotes are indicated thus: 32n

A

acceleration, 7–9, 38
acceptance. *see* taking over
'accepted as noted', definition, 252
'accepted', definition, 252
access to site, 10–11, 194–5
achieved milestone, definition, 252
additional equipment, claims, 39
additional work. *see also* dayworks; extra works
  claims, 39
adjudication, 14–15
administrative receivership, 144–5, 148n
advance-payment bonds, 29
affidavit, 101n, 153n
affiliate, definition, 252
agent bank, definition, 252
'agreement in principle'. *see* letters of intent
Alternative Dispute Resolution (ADR), 12–18. *see also* arbitration
ambiguous clauses, 19–20, 39. *see also* contra proferentem rule; contract maxims and rules of interpretation; implied terms
applicable laws and regulations, 21–3
  definition, 253, 267n
approvals. *see also* consents and approvals, government approvals
arbitration. *see also* Alternative Dispute Resolution (ADR)
  claims, 51–3
assignment, 24–7
authority, definition, 253
award date, definition, 253

B

back charges, 133–4, 136n
bankruptcy. *see* insolvency
banks
  letters of credit, 162–4
banks *(continued)*
  set-off, 226
bid. *see* tender bid
bonds
  additional costs, 39
  general, 29–31
  indemnity, 141
  retention, 212
breach of contract. *see* claims; damages; losses
'burden of proof', claims, 39
business day, definition, 253

C

cause (consideration), 58–9
Certificate of Final Completion, definition, 253
change in law, definition, 253–4
change in the works. *see also* additional work; extra works; material change; omission of work
  definition, 254
change orders
  definition, 254
  management and control, 33–7, 130
  model proposal, 299
'chose in action', 27n
claims
  arbitration, 51–3
  consistency, 39
  definition, 254
  errors in, 116–17
  management, 130–3
  prompt list, 38–45
  review and preparation, 46–50
claims (insurance). *see* insurance, claims
close out, 66–8
  check list, 303–4
comity, 148n
competent authority, definition, 254–5
completion
  Certificate of Final Completion, definition, 253

completion *(continued)*
contract close out, 66–8
date for completion, definition, 256
date of completion, definition, 256
definition, 255
final completion, definition, 267n
Notice of Completion, definition, 264
conciliation, 13
concurrent delay, 39, 54–6
'conditions', 140n
conflicts (ambiguities). *see* ambiguous clauses
consents and approvals, definition, 255
consequential losses, 39, 76–80
'consideration', 57–60
consistency of claims, 39
constructive acceleration, 7–9
constructive variations, claims, 40, 45n
consumables, claims, 40
contemporary records, 61–3
contra proferentem rule, 39, 64–5, 71
contract close out, 66–8
check list, 303–4
contract management, 127–36. *see also* change orders, management and control
contract maxims and rules of interpretation, 69–74
contract price, definition, 255
'contract to be read as a whole', 69
contractor, definition, 255
contractor good industry practice, definition, 256
contractor's duty to warn, 105–8
contractor's equipment, definition, 255–6
contractor's information, definition, 256
contractor's key personnel, definition, 256
contractor's pre-tender information confirmation, 289–90
contractor's pre-tender site review, prompt list, 279–87
Contracts (Applicable Law) Act 1990, 21
copyright, 267n
correspondence, records, 103. *see also* Request for Information (RFI)
costs. *see also* payment
analysis for claims, 48
claims prompt list, 38–45
dayworks, 81–3
definition, 256
records, 103
Counterpart Execution, 59–60
country, definition, 256
currency foreign/local, definitions, 260, 263

D

damages, 40, 52, 75–80. *see also* claims
date for completion, definition, 256
date of completion, definition, 256
day, definition, 257
dayworks, 35, 81–3
record sheet, 291
decennial liability, 84–5
default (conditional) bonds, 30
default, definition, 257
defects. *see also* fitness for purpose
contractor's duty to warn, 105–8
decennial liability, 84–5
definition, 257
liability period, 159–61
deficiency, definition, 257
delays
access to site, 10
analysis, 86–90
concurrent delay, 39, 54–6

delays *(continued)*
employer caused delay, definition, 258–9
float ownership, 122–4
mitigation, 174–6
time at large, 238–41
time for review, 242–3
'time is of the essence', 244–6
deliveries, control, 134
delivery schedule, definition, 257
depositions, 91–6. *see also* discovery; interrogatories
depreciation costs, claims, 40
direct costs, claims, 40
direct labour costs, claims, 40
direct material costs, claims, 40
directed acceleration, 7–9
direction, definition, 257
disclosure. *see* depositions; discovery
discovery, 97–101
discrepancies. *see* ambiguous clauses
Dispute Review Boards (DRBs), 17
disputes. *see also* Alternative Dispute Resolution (ADR)
definition, 258
disruption of work
claims, 43
delay analysis, 86
dissolution, definition, 257–8
documents. *see also* change orders; claims; notices; records, management of
control, 102–4, 128
definition, 258
discovery, 97–101
minutes of meetings, 171–3
monthly progress reports, 177–9
time for review, 242–3
drawings and specifications, definition, 258
duties. *see* import duties
duty to warn, 105–8

E

early neutral evaluation, 15–16
effective date, definition, 258
efficiency loss, claims, 42
efficiency reductions, claims, 40–1
Eichleay's formula, 50n
'ejusdem generis', 71
embargoes (trade), 208
Emben's formula, 50n
employer caused delay, 7–9, 10, 54–6, 194–5, 238–41
definition, 258–9
employer, definition, 258
employer's representative, definition, 259
equipment, contractor's, definition, 255–6
equipment costs, claims, 41
'equity', 245–6n
errors
in claims, 116–17
in tender rates, 235–7
escalation costs, claims, 41
estoppel, 109–11
'examination for discovery', 101n
exclusion clauses. *see* exemption clauses
exclusive remedies, 112–13
exemption clauses
contra proferentem rule, 64–5
force majeure, 125
general, 114–18
rights of third parties, 216
set-off, 226–7
sub-soil reports, 233–4
expert, definition, 259
expert determination, 15
expert witness, 92, 94–5
'expressio unius', 20n, 70–1
expression of interest. *see* letters of intent

extension of time clauses, 238–41, 244–6
extra cost, definition, 259
extra works. *see also* additional work
  claims, 41
  pricing, 36

F

facility, definition, 259
final account stage. *see* close out
Final Completion Certificate, 67, 305
  definition, 267n
final completion date, definition, 259–60
final completion, definition, 267n
finance agreements, definition, 260
finance charges, claims, 41
financial close, definition, 260
fitness for purpose, 119–21
float (schedules), ownership, 122–4
'force majeure', 125–6
foreign currency, definition, 260
forfeiture clauses, contra proferentem rule, 64
'fragnet', 50n
fraudulent claims, 116–17

G

glossary, 251–67
golden rule, 72
good operating practices, definition, 256, 260
government approvals, definition, 260
government authority, definition, 260
guarantees, 29–31, 142–3

H

Hague Convention, 21–2
hazardous materials, definition, 261
heads of agreement. *see* letters of intent
home office overheads, claims, 41
Housing Grants, Construction and Regeneration Act 1996 (HGCRA), 14
Hudson's formula, 50n

I

implied terms, 137–40
import duties, 134–5, 208–9
  definition, 261
inconsistencies. *see* ambiguous clauses
indemnity, 141–3
Independent Engineer. *see* Lender's Engineer
indirect costs, claims, 41
indirect losses, 76–80
information. *see also* contractor's information; documents; records, management of; Request for Information (RFI)
  definition, 261
insolvency. *see also* guarantees
  definition, 261
  general, 144–8
  set-off, 226
inspections
  access to site, 11
  quality control, 135
insurance. *see also* decennial liability
  claims, 149–51
  costs, claims, 42
  exclusive remedies, 113
  indemnity, 141
  subrogation, 231–2
intellectual property rights, definition, 261–2
'intention is to be found from the contract itself', 69
'interested parties', definition, 262
'intermediate (or innominate term)', 140n
international legal systems, 269–78
interrogatories, 98, 101n, 152–3

J

job site. *see* site

K

kick-off meetings, 155–7

L

language, ruling, 223–4
late payment rate, definition, 262
latent defects, liability period, 159–61
law. *see also* international legal systems
  applicable laws and regulations, 21–3
  changes, definition, 253–4
  contract maxims and rules of interpretation, 69–74
  definition, 262
law report abbreviations & table of cases, 307–13
legal systems, international, 269–78
lenders, definition, 262
lender's engineer, definition, 262
letters of awareness. *see* letters of intent
letters of credit, 162–4
letters of intent, 165–9
liabilities. *see* claims; damages
liability, definition, 263
liquidated damages, 75
  contra proferentem rule, 64
liquidation, 145–6
literal rule, 72
local currency, definition, 263
loss of efficiency, claims, 42
losses. *see also* claims; damages
  claims prompt list, 40

M

major sub-contract, definition, 263
management
  change orders, 33–7
  claims, 130–3
management *(continued)*
  contracts, 127–36
  records, 44, 61–3
material change, definition, 263
material costs, claims, 42
materials, control, 134–5
materials, plant and equipment, definition, 263
med/arb, 13–14
mediation, 13
meetings. *see also* kick-off meetings
  minutes, 171–3
memoranda of understanding/intent. *see* letters of intent
milestone, definition, 252, 264
milestone payment, definition, 264
mini-trials, 16
minutes of meetings, 171–3
mischief rule, 72
mitigation, 42, 174–6
month, definition, 264
monthly progress reports, 177–9
  format, 295–8

N

negligence, liability, 116, 160
negotiation procedures, 181–5
  'subject to contract', 228–30
  'without prejudice', 247–8
New York law, 23n
notice of completion, definition, 264
notice to proceed, definition, 264
notices, 42, 186–8
  compliance, 129
  definition, 264
novation, 25–7

O

'obligation de moyens', 201
'obligation de resultat', 201–2
office overheads, claims, 41

omission of work, loss of contractor's profit, 189–90
on-demand (unconditional) bonds, 30
'ordinary meaning', 69
OS&D reports, 179n
out of sequence working, claims, 42
Overage, Shortage and Damage (OSD) reports, 179n, 293
overhead costs, claims, 43
overtime costs, claims, 43

P

'parastatal', 210, 211n
'particulars prevail over standards', 70
patents, 267n
payment. *see also* 'consideration'
  dayworks, 81–3
  late payment rate, definition, 262
  letters of credit, 162–4
  letters of intent, 165–9
  milestone payment, definition, 264
  quantum meruit, 197–200
  retention, 212–13
  set-off, 225–7
  tender rate errors, 235–7
performance bonds, 30
permanent work, definition, 264
permit, definition, 264–5
person, definition, 265
plant, definition, 263
pre-tender information confirmation, 289–90
pre-tender information review, 191–3
pre-tender site review, prompt list, 279–87
presumption, claims, 43
prevention, 194–5
  time at large, 238–41
privilege (discovery defence), 97, 99
'privity of contract', 216n
procedural law, 23n
profit margin
  claims, 43
  loss due to omission of work, 189–91
programmes (project). *see* schedules
protocols. *see* letters of intent
proving entitlement, claims, 43
provisional acceptance, 66. *see also* contract close out
  definition, 267n
Provisional Acceptance Certificate, definition, 267n
provisional acceptance date, definition, 265
punch list, definition, 265–6
purposive approach, 72–3

Q

quality control, 104, 135–6
quantification, damages, 52
'quantum meruit', 198–200
quasi-contracts, 198, 200n
quotas (trade), 208

R

'reasonable skill and care', 201–2
receivership, 144–5
recognising claims, 44
records. *see also* documents
  management of, 44, 61–3, 102–4
  research for claims, 47–8
regulations. *see* applicable laws and regulations
rejection, claims, 44
relationship management, 129
remedial work, access to site, 11
reports, monthly, 177–9
Request for Information (RFI), 34, 203–4
Request for Proposal (RFP), 34–5
restitution, 205–7
restricted access, claims, 44

retention (payments), 212–13
bonds, 30
ripple effects, claims, 44, 45n
risks
allocation, assignment and assumption, 217–22
contractor's duty to warn, 105–8
indemnity, 141–3
mitigation, 174–6
pre-tender review, 191–3
Rome Convention, 21–2
rules of construction, 19, 73
ruling language, 223–4

S

schedules. *see also* delivery schedule
concurrent delay, 54–6
delay analysis, 48, 86–90
float ownership, 122–4
management, 133
records, 104
set-off, 225–7
'simple' contracts, 60n
site. *see also* access to site
definition, 262, 266
overheads, claims, 42
pre-tender information review, 191–3, 279–87
specifications. *see* drawings and specifications
standard forms of contract, 65n
standards of work, 201–2. *see also* good operating practice
statutory requirement, definition, 266
sub-contract close-out, 66–7
check list, 303–4
sub-contractors
control, 133–4
definition, 266
Final Completion Certificate, 305
kick-off meeting, 155–7
retention of payment, 212–13
sub-contractors *(continued)*
Taking Over Certificate, 301
sub-soil, investigations, 233–4
'subject to contract', 228–30
subrogation, 231–2
subsidies (trade), 208
substantial completion, 66. *see also* contract close out
substantive law, 23n
sufficiency of resources, claims, 44
suppliers, control, 133–4

T

taking over, 66
Taking Over Certificate, 301
tariffs (trade), 134–5, 208–9
tax, definition, 266
temporary work, definition, 266
tender bids
documents, 102
pre-tender information review, 191–3
tender bonds, 29
tender rate errors, 235–7
term sheets. *see* letters of intent
third parties, rights of, 214–16
time at large, 238–41
time for review, 242–3
'time is of the essence', 244–6
trade marks, 267n
trade restrictions, 208–11

U

un-liquidated damages, 75
unfair contracts, 114–15
unjust enrichment, restitution, 205–7

V

'validity of meaning', 70
valuable consideration. *see* 'consideration'

W

warranty, 140n
warranty bonds, 30
'without prejudice', 247–8
witness statements, 62. *see also* depositions; discovery
'words will prevail over figures', 70
work rescheduling, claims, 45
worker, definition, 267
work(s), definition, 266–7